Environmental Impact of Chemicals:
Assessment and Control

Environmental Impact of Chemicals: Assessment and Control

Edited by

Michael D. Quint

Dames & Moore, Twickenham, UK

David Taylor

Brixham Environmental Laboratory, Zeneca Limited, Brixham, UK

Rupert Purchase

Environmental Chemistry Group, The Royal Society of Chemistry, London, UK

THE ROYAL
SOCIETY OF
CHEMISTRY
Information
Services

Based on the Proceedings of a Symposium on Acquiring Environmental Data for Legislative Needs, organized jointly by the RSC Environmental Chemistry and Chemical Information Subject Groups, held on 17 October 1994 and a Symposium on Toxicology and Quantitative Risk Assessment, organized by the RSC Toxicology Subject Group (in Association with the Health and Safety Group of the Society of Chemical Industry), held on 9 November 1994.

Special Publication No. 176

ISBN 0–85404–795–6

A catalogue record of this book is available from the British Library.

Published by The Royal Society of Chemistry,
Thomas Graham House, Science Park, Milton Road, Cambridge CB4 4WF, UK

Typeset by Vision Typesetting, Manchester
Printed in Great Britain by Hartnolls Ltd., Bodmin, UK

Preface

There is little doubt that industrial chemicals are capable of causing adverse environmental impacts. These can arise through their deliberate use, as pesticides for example, or due to their planned or unplanned release from manufacturing processes.

The traditional approach to dealing with such impacts has tended to be of a reactive nature, whereby control has been delayed until cause and effect are proven. This is being replaced, however, by a more proactive approach, in which impacts are predicted and, if necessary, prevented, *before* they actually occur. The driving force behind this change is a broad acceptance of the "precautionary principle", which is defined in the Rio Declaration as follows:*

> *"Where there are threats of serious or irreversible damage, lack of full scientific certainty shall not be used as a reason for postponing cost-effective measures to prevent environmental degradation."*

The aim of this book is to provide an insight into a technique that is being used to help predict the environmental impact of chemicals in specific circumstances, namely risk assessment. The tools of environmental risk assessment – toxicology, epidemiology, exposure modelling and analytical chemistry – are outlined, along with the means for applying them within a regulatory framework.

Following an introduction and overview of the risk assessment process (Chapter 1), the roles of toxicology and epidemiology are discussed (Chapters 2–6), with views from Europe and the United States. Recent environmental legislation in the UK has focused on the need to protect the environment within reasonable economic constraints; the role of risk assessment in this is examined (Chapters 7–10).

The importance of incorporating site-specific data into risk assessments is described (Chapter 11) along with the collection of information on new and existing chemical substances, with particular regard to recent European Union legislation (Chapters 12–14). The book concludes with a discussion of the interplay between environmental risk assessment and the realities of public perception (Chapters 15 and 16).

* 'A Guide to Risk Assessment and Risk Management for Environmental Protection', Department of the Environment, HMSO, 1995.

The book is based on the proceedings of two symposia held in London in October and November, 1994 and organized jointly by the Toxicology, Environmental Chemistry and Chemical Information Subject Groups of the Royal Society of Chemistry, in association with the Health and Safety Group of the Society of Chemical Industry. We thank the committee members of these respective groups for their support.

Michael Quint, David Taylor and Rupert Purchase
London, January 1996

Contents

Chapter 1

Overview of Risk Assessment and Its Application

By Andrew Jackson and Gev Eduljee

ENVIRONMENTAL RESOURCES MANAGEMENT, EATON HOUSE,
WALLBROOK COURT, NORTH HINKSEY LANE,
OXFORD OX2 0QS, UK

1   Introduction

The application of risk assessment in the UK has historically been associated with the technologically based major hazard and safety fields: indeed one commentator states that "the discipline of risk assessment should be seen as a branch of engineering".[1] The use of this technique to analyse chronic (as opposed to catastrophic) releases, however, and their related public health effects, is relatively new in the UK, outside of the radiological field. The further application of risk assessment to inform policy decisions across a wide spectrum of regulatory activities is even more recent, dating from the early 1990s. In this respect, the US and some countries in continental Europe, notably the Netherlands, have developed formalized risk assessment and risk management procedures since the early 1980s, and have incorporated this approach into national policy-making to a far greater extent than is the case in the UK.

This chapter examines the potential for application of risk assessment in the UK, drawing on the experiences in the US and the Netherlands. The discussion commences with an overview of the risk assessment and risk management process, and then outlines current initiatives in the UK in which this approach is being applied and tested. Three case studies are then presented to illustrate various aspects of the methodology. Finally, we offer some thoughts as to the directions in which risk assessment could be further developed in the UK over the coming years.

2   The Risk Management Process and Its Application

Risk Assessment and Risk Management

Risk *assessment* denotes the process of estimating and evaluating the risk posed by a particular activity or exposure. On its own, risk assessment has limited value. It is within the overall framework of risk *management* that the loop representing the decision-making process is closed. Risk management has been defined as "the process whereby decisions are made to accept a known or assessed risk and/or the implementation of actions to reduce the consequences or probabilities of occurrence".[2] The risk management process is generally described as comprising four or five stages, depending on the particular regulatory background. For example, the stages may be as follows:[3]

- Hazard identification
- Hazard analysis
- Risk estimation
- Risk evaluation
- Risk control

The risk management policy of the Dutch government is also a five-stage process, but with the following descriptors:[4]

- Hazard identification
- Quantification of risk
- Decision-making
- Risk reduction
- Risk control

Regardless of the specific terms used, the objectives and the process of assessing and managing risk are comparable. The process is iterative in that having identified and quantified the risk, the measures that are proposed to control or reduce the risk are taken back through the risk assessment procedure to evaluate their effectiveness, both in relation to the cost of these measures relative to their benefits, and in relation to their technical efficacy. Figure 1 illustrates the concept.[5]

The distinction between *assessment* and *management* of risk is a key issue. The purpose of risk assessment should be to evaluate, as accurately as possible, the nature and magnitude of a potential risk, and equally importantly, to state what

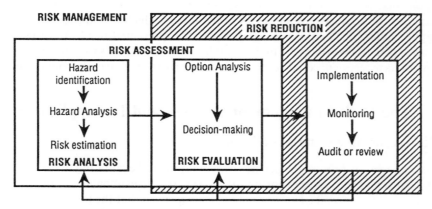

Figure 1   *The risk management process*

is not known about the risk (*e.g.* the uncertainty associated with a toxicity assessment). The function of risk management is to decide whether this level of risk is acceptable and, if not, to translate this information into actions and measures designed to control exposure and/or to reduce risk in other ways. In the US it has been argued that the Environmental Protection Agency (EPA) has allowed a blurring of this distinction by insisting that excessively conservative assumptions be built into the risk assessment procedure, displacing the role of public protection from its proper position within the risk management stage. Whereas the use of conservative assumptions is seen as a legitimate means of allowing for uncertainty, the preferred approach is to balance the worst-case (or high-end) risk with a central estimate of risk, in order to place the former in perspective.

Risk Assessment as Applied in the US and the Netherlands

The US and the Netherlands are taken as examples of countries in which risk assessment has been progressively integrated into national environmental policy- and decision-making over the past two decades. While risk assessment is always implicit in the decision-making process, these countries have recognized the benefit of formalizing the analysis of risk.

The Netherlands introduced the concept of risk assessment into environmental policy-making in their 1986–1990 Programme for Environmental Management, formally appending a policy of systematic risk management to the Dutch Environmental Policy Plan in 1990.[6] Two types of environmental measures were identified:

- *Source-related* policies, concerned with the prevention of environmental pollution by controlling releases at source.
- *Effects-oriented* policies, which address the effects of environmental pollution on people and the ecosystem.

The strength of this approach lies in the complementary nature of the two types of policies. Risk assessment is applied in an effects-oriented approach to determine the effect of pollutants on exposed environmental receptors and to establish suitable standards for environmental protection. These standards are then translated into source-oriented policies to control the releases of the pollutants such that the environmental standards are not breached. Risk management was seen as a means to ensure comparability of protection levels and methods of dealing with pollution sources, and to harmonize the setting of standards in the various environmental media, thereby creating a coherent environmental policy.

An interesting application of this approach has been in the planning field, where the concept of risk assessment has been used to establish qualitative acceptability criteria for different types of effects.[7] A zoning system called 'Integral Environmental Zoning' (IEZ) has been developed, in which five classes of environmental quality (A to E in ascending order of pollution load) have been defined for various types of hazard (see Table 1). The environmental effect of, say,

Table 1 Classification of the IEZ sectoral environmental loads for the maximum permissible and the negligible levels for existing and new situations[7]

Agent	Class E	Class D	Class C	Class B	Class A
Existing situations (Class A has no significance)					
Individual risk (per year)	$>10^{-5}$	10^{-5}–10^{-6}	10^{-6}–10^{-7}	$<10^{-7}$	($<10^{-8}$)
Noise [dB(A)]	>65	65–60	60–55	<55	(<50)
Odour (OU m^{-3})	>10	10–3	3–1	<1	(<1 [99.5%])
Carcinogenic substances (risk per year)	$>10^{-5}$	10^{-5}–10^{-6}	10^{-6}–10^{-7}	$<10^{-7}$	($<10^{-8}$)
Toxic substances (Fraction of No Effect Level, NEL)	$>1 \times$ NEL	0.1–$1 \times$ NEL	0.03–$0.1 \times$ NEL	$<0.03 \times$ NEL	$<0.01 \times$ NEL
New situations (Class E has no significance)					
Individual risk (per year)	($>10^{-5}$)	$>10^{-6}$	10^{-6}–10^{-7}	10^{-7}–10^{-8}	$<10^{-8}$
Noise [dB(A)]	(>65)	>60	60–55	55–60	<50
Odour (OU m^{-3})	(>10)	>3	3–1	<1	<1 (99.5%)
Carcinogenic substances (risk per year)	($>10^{-5}$)	$>10^{-6}$	10^{-6}–10^{-7}	10^{-7}–10^{-8}	$<10^{-8}$
Toxic substances (Fraction of No Effect Level, NEL)	($> \times 1$ NEL)	$>0.1 \times$ NEL	0.03–$0.1 \times$ NEL	0.01–$0.03 \times$ NEL	$<0.01 \times$ NEL

a 60 dB(A) noise level in Class D has the same severity as 1 Odour Unit m^{-3} in Class D. An integrated environmental index is then calculated to assess the combined effects of different types of environmental effects.

Most national environmental policies, though not explicitly stated, contain an amalgam of source- and effect-oriented approaches. For example, the US Clean Air Act, amended in 1990, requires the federal body charged with the implementation of environmental policy, the EPA, to impose technology-driven emission standards on industry, and in addition set residual risk standards to protect public health.

In the US, risk assessment has long been a pivotal element in decision-making, largely on a site-specific basis in relation to the management of contaminated land, but latterly extended to encompass the scrutiny of policy decisions. For example, the EPA is required to ensure a balance between the risks to man and the environment from the use of registered pesticides, and the economic, social, and environmental costs and benefits accruing from their use. In a nationwide exercise, the EPA is assisting individual states to undertake comparative risk analysis of a range of environmental issues, in order to set environmental priorities. The initiative is unique in that public participation in the decision-making process is being sought, so as to avoid the possibility of generating confrontational and non-representational policies that have no public support. The results of the exercise will be fed into the EPA's national strategic plan.

Other countries have increasingly integrated risk assessment into specific areas of policy. The management of contaminated land is perhaps the most common environmental issue that has been regulated using the principles of risk assessment. Canada, Australia, Germany, New Zealand, and Austria are included in those countries that have followed the US and the Netherlands in developing national strategies and site-specific risk assessment protocols for this purpose.

3 Risk Assessment and Risk Management in the UK

As noted in Section 1, risk assessment has mainly been used in the UK to manage the environmental effects of major accidents and of radioactive substances, and to regulate the sources of these hazards or releases. This section examines the more general use of risk assessment in the environmental arena.

Risk Assessment of Waste Management Options

The formal process of risk assessment has been used to evaluate the impact of waste management activities, especially of emissions from combustion processes, and of releases from landfills. The interest in a more formal and sophisticated evaluation of waste management processes coincided with the introduction of European Union (EU) legislation requiring environmental assessments to be performed as part of the planning procedure for such facilities. Assessments have been primarily undertaken by developers and operators of waste disposal facilities, using assessment techniques based on the US EPA Superfund and

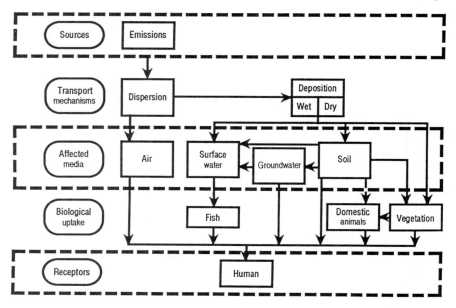

Figure 2   *Exposure pathways following emissions to atmosphere*

Human Exposure Models. Figure 2 illustrates the assessment procedure for evaluating the impact of combustion emissions released to atmosphere. Two pathways are assessed:

- The direct uptake of chemicals from the stack emissions, via inhalation.
- The indirect uptake of chemicals, following deposition of stack emissions onto soil, surface waters, and vegetation, and transfer through the foodchain to humans.

The risk assessment framework can provide useful information on potential risk management options. For example, for a given release from a proposed plant, a target risk level for a receptor in the receiving environment can be met by either lowering the mass of the chemical released over time (by improving the design of the abatement system) or by increasing the stack height and emission velocity of the release to create more efficient dispersion. Figure 3 shows the relationship between stack height and stack exit velocity and the resulting ambient air concentration in the environment, for a hypothetical release. The selection of one or other of the options is facilitated by the systematic assessment and analysis of environmental effects, mitigation measures and their associated costs.

The Department of the Environment (DoE) has funded risk-based research into landfill design and operation for many years. A risk assessment protocol has been developed to study the health effects of hazardous wastes in landfills, and classical risk assessment techniques have been applied to problems such as predicting the failure rate of liner systems. Risk assessment is especially suited to the assessment of landfill operations, since the release of leachate and potential

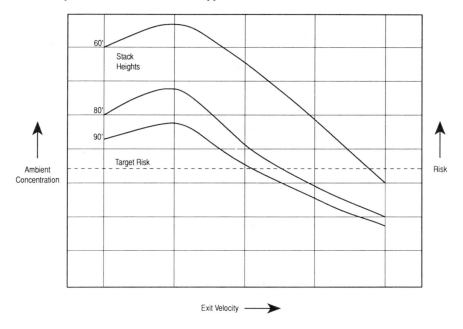

Figure 3 *Risk vs. exit velocity and stack height (feet) for fixed temperatures*

contamination of groundwater occurs over long timescales: risk assessment predicts the time-dependent movement of a pollution plume, starting from current operations and extending the prediction decades into the future. The efficacy of various design features and mitigation measures can be taken through the methodology in order to predict the long-term effects on the environment.

The Management of Contaminated Land

The management of contaminated land has been one of the first environmental issues to be systematically addressed by the DoE, using the risk assessment approach.[8] Following the adoption of a risk assessment framework for the management of contaminated land in 1991, the Department has continued to develop the necessary tools to predict the behaviour of pollutants in various environmental media, to develop criteria for the clean-up and protection of soil and water, and to evaluate the relative costs and benefits of remedial technologies. An example of how the risk management paradigm illustrated in Figure 1 can be used as the basis for establishing research needs is provided by the projects funded by the Department since 1991.

- *Hazard identification*: the development of industry profiles, characterizing the types of operations that could lead to contamination and the chemicals encountered in contaminated land.

- *Hazard analysis*: the development of a computer model, Contaminated Land Exposure Assessment (CLEA) to assess human exposure from contaminated land.[9]
- *Risk estimation* and *risk evaluation*: compilation of toxicological profiles of the chemicals of interest; this information will then be merged with CLEA to facilitate risk evaluation.
- *Risk control*: evaluation of remedial techniques, cost-benefit analysis of remediation technologies, long-term performance of treatment methods, etc.

Central government is not alone in applying risk assessment to the problem of contaminated land. For example, the Welsh Development Agency has published a framework for risk assessment which mirrors the US Superfund approach, and provides a consistent basis for assessing contamination in the Principality.[10] Other local authority and regional development agencies have also expressed a preference for this approach.

The Management of Environmental Pollution

Several other examples of the application of risk assessment in the regulatory field can be cited. For example, the National Rivers Authority (NRA) has funded joint research with the DoE on PRAIRIE, a risk assessment system to assess the Pollution Risk from Accidental Influxes to Rivers and Estuaries. NRA's interest clearly overlaps with the DoE's in the area of contaminated land and landfills, since water resources could be at risk from such sites. Joint activities with the Department are in progress to set priorities for clean-up of contaminated land and to assess the impact of landfill leachate on groundwater resources.

Her Majesty's Inspectorate of Pollution (HMIP) has reviewed the potential for applying risk assessment and risk management techniques to assist in performing its regulatory duties.[11] The following were identified as areas of activity that were amenable to the application of risk management techniques:

- Assessing applications for authorization under Integrated Pollution Control (IPC), whereby risk-based environmental indices would be developed alongside the current hazard-based Best Practicable Environmental Option (BPEO) assessment system.[12]
- Setting policy priorities based on risk. In this respect it was recommended that HMIP study the US EPA's approach to policy appraisal (see Risk Assessment as Applied in the US and the Netherlands Section).
- Setting regional and national environmental priorities, again based on the US EPA's comparative risk approach.
- Prioritizing inspection programmes, building on existing qualitative risk scoring techniques used by field Inspectors.

HMIP has also created the Centre for Integrated Environmental Risk Assessment (CIERA) as part of its strategy to develop expertise in the assessment and management of environmental risks from industrial activities. The Centre has been formed within the Integrated Environmental Management Branch of HMIP's Pollution Policy Division, and draws on expertise both from within HMIP and from outside of Government.

The DoE has published a consultation document describing the overall Government strategy to reduce the emissions of hazardous chemicals to the environment.[13] The document was part of Government's intention to control diffuse sources of hazardous chemicals and to encourage the use of safer ones. In the document, risk assessment was considered central to identification of substances that are likely to be dangerous when released into the environment, evaluation of their effects when released, and appraising various methods of control.

The DoE has recently published *A Guide to Risk Assessment and Risk Management for Environmental Protection*, which sets out a formal risk management framework within which the diverse activities of the Department can be accommodated. Risk management is seen as comprising two parts: risk assessment (comprising risk estimation and risk evaluation) and risk management. Risk estimation has five stages:

- Stage 1: Description of intention
- Stage 2: Hazard identification
- Stage 3: Identification of consequences
- Stage 4: Estimation of magnitude of consequences
- Stage 5: Estimation of probability of consequences

In the second stage of risk assessment, the estimated risk is evaluated for significance.

Finally, risk management is defined as the process of implementing decisions about tolerability or altering risks. Monitoring of the outcome of a project forms an integral part of the process. The methodology is now being tested by the Department. A particular challenge will be to integrate the new methodology into existing risk management systems, given that the terminology will be different.

4 Case Studies

This section presents three short case studies to demonstrate typical examples of the current application of risk assessment in the UK.

Case Study 1: Assessment of Releases from a Secondary Lead Smelter

Releases of lead into the environment have attracted considerable academic and public attention for over 20 years and this is mirrored by the efforts made by regulators. One of the many manifestations of the level of interest in this pollutant is a quite extensive body of data, some of which reports background concentrations in a variety of environmental media and some of which is related to particular facilities. This case study summarizes the evaluation of an extensive set of monitoring data for a secondary lead smelter.[14]

A risk assessment was undertaken to determine the principal sources of human exposure to lead releases from the facility. The Uptake Biokinetic Model (UBK) was used to determine the relative exposures from the following sources:

- air
- diet
- potable waters
- indoor dust
- soil and outdoor dust

Blood lead concentrations were used as the key indicator of lead intakes and the predicted concentrations were compared with the following:

- the UK blood lead action level of $25 \mu g \ dl^{-1}$, as a means of assessing the significance of exposures;
- blood lead monitoring surveys in the area, as a means of assessing the accuracy of the UBK model.

Predicted concentrations of lead in blood at five locations in the vicinity of the smelter are presented in Figure 4 as probability density functions. It is evident, that at one of the locations, exposures were estimated to be elevated, resulting in an estimated 4.5% of the exposed population having blood lead levels greater than the action level of $25 \mu g \ dl^{-1}$. The geometric mean predicted blood lead concentration of $14.1 \mu g \ dl^{-1}$ compared well with the observed value of $13.6 \mu g \ dl^{-1}$.

An examination of the relative significance of exposure pathways is summarized in Table 2. It is evident that the ingestion of soils and dusts is the most significant pathway and that inhalation is of the least significance. Should it be considered that lead exposures in the local population are excessive and should be reduced, then attention should primarily be paid to reducing exposure to contaminated soils and dusts. The existing inventory of lead in soils is very large when compared with on-going deposition as a result of stack gas emissions from the smelter. Fugitive emissions from the facility may be contributing to the inventory of lead in soils and dusts, for example, vehicles leaving the site may entrain contaminated dust on their wheels which is subsequently deposited on the roads. Hence, the priority is to examine the potential for such modes of release and their control rather than stack gas emissions.

Case Study 2: The Use of Sewage Sludge in Agriculture as a Source of Exposure to Dioxins and Furans

Whereas Case Study 1 demonstrated the use of risk assessment for the retrospective analysis of the operation of a particular facility, this Case Study

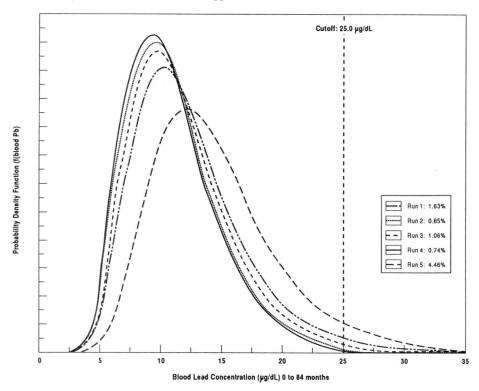

Figure 4 *Distribution of blood lead concentrations*

Table 2 *Sources of lead exposure*

Age group	Mean blood level (μg dl^{-1})	Uptake (μg day^{-1})				
		Total	*Soil and dust*	*Diet*	*Water*	*Air*
0.5 to 1	13.5	45.7	41	2.9	1.7	0.1
1 to 2	14.1	48.4	41	3.0	4.4	0.1
2 to 3	14.0	49.1	41	3.4	4.5	0.2
3 to 4	14.2	49.2	41	3.3	4.6	0.3
4 to 5	14.8	49.2	41	3.2	4.8	0.3
5 to 6	14.8	49.9	41	3.4	5.1	0.5
6 to 7	14.7	50.2	41	3.7	5.1	0.3

examines the potential implications of a change in current practice for the disposal of sewage sludge. It is anticipated that the implementation of the EC Directive on urban waste water treatment will increase the quantities of sludge produced and remove the option of sea disposal, thereby increasing the quantities of sludge to be applied to agricultural land. The UK DoE has produced evidence indicating the presence of polychlorinated dibenzo-*p*-dioxins and dibenzofurans in sewage sludges and concerns have been expressed in some quarters that human exposures may be increased to an unacceptable level.

In order to address this matter an extreme worst-case scenario was developed to assess exposures via the food chain following the application of sludge to agricultural land.[15] The aim was to produce an upper bound estimate of exposure to dioxins and furans in the form of International Toxic Equivalents (I-TEQ) for comparison with the UK Tolerable Daily Intake (TDI) of 10 pg I-TEQ kg^{-1} day^{-1} and the estimated background exposure of approximately 2 pg I-TEQ kg^{-1} day. The exposure pathways addressed are summarized in Figure 5. The assumed concentration of dioxins and furans in the sludge was varied according to the four catchment types identified by the DoE:

- rural, 23.3 ng I-TEQ kg^{-1} dwt
- mixed industrial/rural, 42.5 ng I-TEQ kg^{-1} dwt
- light industrial/domestic, 42.3 ng I-TEQ kg^{-1} dwt
- industrial/domestic, 52.8 ng I-TEQ kg^{-1} dwt

The principal sources of exposure were predicted to be the ingestion of beef (58%) and milk and dairy products (29%). Hence, the selective use of sewage

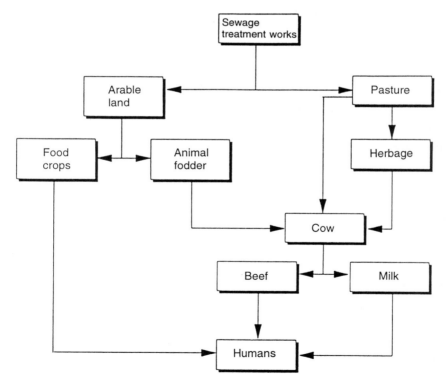

Figure 5   *Assessment framework for the application of sewage sludge to land*

sludge for application to arable land, rather than pasture, would result in a significant reduction in intakes.

The estimated incremental daily intakes ranged from 0.05 to 0.11 pg kg^{-1} day^{-1} and from 0.36 to 0.80 pg kg^{-1} day^{-1} for soils receiving single and 10 applications respectively. Taking the highest estimated exposure, this approximates to a 40% increase in background exposures but still falls well below the TDI.

Case Study 3: Assessment of Risks from a Closed Landfill Site

In common with many industrialized countries, the UK has a legacy of closed landfill sites requiring informed management to ensure that they do not pose a risk to human health and the environment. Many of these sites were designed and operated before the advent of engineered containment sites which are currently favoured. The use of risk assessment enables the most significant exposure pathways presented by a closed landfill site to be identified, thereby acting as a guide for any subsequent risk management measures considered appropriate. This case study exemplifies such an assessment.

The site under consideration had received a variety of waste types over its operational lifetime, including a number of industrial wastes. A detailed sampling and analytical programme had been undertaken with a view to establishing the nature and extent of contamination present in the waste mass, leachate, and neighbouring water bodies. The site has a complex hydrogeology which exacerbates problems created due to the lack of a liner to control the migration of leachate.

Exposures were calculated for three groups in the population:

- on-site trespassers
- neighbouring residents
- persons infrequently using the site and surrounding area for recreational purposes

Carcinogenic risks and non-carcinogenic hazards were calculated using toxicological data from the US EPA's Integrated Risk Information System. A summary of the risks and hazards posed by the various pathways examined is presented in Table 3.

A lifetime risk of one in a million (or 1×10^{-6}) was, for the purposes of this assessment, considered to be the level at which the consideration of remedial actions was triggered. It was evident that the most significant exposure pathways for carcinogenic risk were direct contact with the enplaced waste mass and the domestic use of groundwater directly below the site. However, the likelihood of these events arising was considered to be extremely low so that as long as preventive measures were put in place, no further action was required.

Non-carcinogenic hazards were addressed with reference to a Hazard Index value of unity. Values below unity were considered to be insignificant, whereas values greater than unity would trigger further consideration. By this criterion, none of the pathways was considered to be of significance.

Table 3   *Summary of risks and hazards*

Route of exposure	Carcinogenic risk	Hazard Index
Direct contact	3.5×10^{-5}	5.5×10^{-3}
Fugitive dust	2.6×10^{-7}	5.3×10^{-3}
Volatization	9.5×10^{-7}	5.0×10^{-2}
Domestic groundwater use	9.1×10^{-5}	2.1×10^{-2}
Groundwater discharge to creek	1.9×10^{-7}	1.9×10^{-4}
Groundwater discharge to river	2.5×10^{-9}	2.5×10^{-6}
Direct leachate release to creek	9.3×10^{-7}	1.3×10^{-1}
Direct leachate release to river	1.2×10^{-8}	1.8×10^{-3}

5   Future Development of Risk Assessment in the UK

The UK is on the threshold of a major change in the organization of its regulatory responsibilities with the formation of the Environment Agency, which will combine HMIP, the NRA, and the local waste regulation authorities. HMIP, through the implementation of IPC and the application of BPEO, currently has to take cross-media considerations into account when regulating industrial processes. Indeed, the requirement for HMIP to apply Best Available Techniques Not Entailing Excessive Cost (BATNEEC) through the Chief Inspector's Guidance Notes, together with the assessment of environmental effects and BPEO, constitutes an excellent example of the Dutch source- and effects-oriented approach to regulatory control (see Risk Assessment as Applied in the US and the Netherlands Section). The risk assessment framework is an ideal vehicle within which to integrate the regulation of industrial releases and cross-media environmental effects.

The need to develop integrated assessment and regulatory systems will assume even greater importance when the current functions of the NRA are set alongside those of HMIP within the Environment Agency. The NRA's joint development of a regulatory approach for contaminated land and water resources will continue. Policy appraisal, priority setting on a national and regional scale, cost benefit analysis, *etc.* will need to be conducted in a consistent manner, for which the risk assessment/management framework is again well suited. This fact is recognized by the various government agencies; for example one of HMIP's objectives in 1995/96 is to develop a corporate risk management policy to guide legislative development. The UK can gain from the experiences of the US and the Netherlands in this arena; CIERA intends to liaise with overseas governments and participate in risk analysis workshops in order to open a dialogue in this area.

The integration of environmental considerations into pollution control and planning control is another area that will benefit from a common approach based on a risk management framework. The DoE's initiative (see The Management of Environmental Pollution Section) is a step in this direction. The Netherlands' use of qualitative risk assessment to assist in the development of locational criteria is an example of how this can be achieved (see Risk Assessment as Applied in the US and the Netherlands Section).

Finally, the UK has a number of international obligations to meet, not least through membership of the EU. The introduction of Integrated Pollution Prevention and Control, international efforts to reduce the releases of environmentally harmful substances, the assessment of the effects of pesticides and genetically modified organisms *etc.* will all require a consistent cross-sector and cross-media approach, wedding regulation of industry and environmental protection with economic assessment and priority setting. The risk assessment and risk management paradigm offers a holistic framework within which to pursue this goal.

References

1 L. E. J. Roberts, in 'Risk Assessment of Chemicals in the Environment', ed. M. L. Richardson, Royal Society of Chemistry, London, 1988.
2 Royal Society, 'Risk: Analysis, Perception and Management', Royal Society, London, 1992.
3 J. Petts and G. Eduljee, 'Environmental Impact Assessment for Waste Treatment and Disposal Facilities', John Wiley, Chichester, UK, 1994.
4 B. J. M. Ale and M. Seaman, in 'Proceedings of the Second European Conference on Environmental Technology', ed. K. J. A. de Waal and W. J. van den Brink, Amsterdam, 22–26 June, 1987, Martinus Nijhoff Publishers, Dordrecht, 1987.
5 S. J. Cox and N. R. S. Tait, 'Reliability, Safety and Risk Management: An Integrated Approach', Butterworth-Heinemann, Oxford, 1991.
6 Dutch National Environmental Policy Plan – Premises for Risk Management, Second Chamber of the States General, Session 1988–1989, 21137, No 5, Ministry of Housing, Physical Planning and Environment, The Hague, 1989.
7 'Manual for a Provisional System of Integrated Environmental Zoning', Ministry of Housing, Physical Planning and Environment, The Hague, 1990.
8 J. Denner, in 'Contaminated Land: Developing a Risk Management Strategy', IBC Technical Services, London, June 1994.
9 C. Ferguson and J. Denner, in 'Contaminated Soil '93', ed. F. Arendt, G. J. Annokkee, R. Bosman and W. J. van den Brink, Kluwer Academic, Dordrecht, 1993.
10 Welsh Development Agency, 'The WDA Manual on The Remediation of Contaminated Land', Welsh Development Agency, Cardiff, 1992.
11 'The Application of Risk Assessment and Risk Management in HMIP's Regulatory Role – Review of the State-of-the-Art', HMIP, London, July 1994.
12 'Environmental, Economic and BPEO Assessment Principles for Integrated Pollution Control', Draft Technical Guidance Note (Environment) E1, Vols. I–III, HMSO, London, 1995.
13 'Reducing Emissions of Hazardous Chemicals to the Environment', Department of the Environment, London, February 1994.
14 'Assessment of the Impact of Lead Releases from Britannia Recycling Limited', HMIP, London, in press.
15 A. P. Jackson and G. H. Eduljee, 'An assessment of the risks associated with PCDDs and PCDFs following the application of sewage sludge to agricultural land in the UK', *Chemosphere*, 1994, **29(12)**, 2523–2543.

Chapter 2

The Role of Toxicology in Risk Assessment

By George Kowalczyk

DAMES & MOORE, BLACKFRIARS HOUSE,
ST. MARY'S PARSONAGE, MANCHESTER, M3 2JA, UK

1 Introduction

The essence of toxicology is to generate information on the biological effects of chemical substances, and to relate that information to potential health consequences for man. The information may be obtained from many sources, but is principally retrieved from animal experiments, and therefore needs to be manipulated, interpreted, and converted into meaningful information in terms of effects upon human health.

Virtually all substances to which we may be exposed, be it in our drinking water, at our workplaces, in the air we breath, or in the food we eat, have been subject to some degree of toxicological investigation. For some substances such as additives in lubricants, we have very limited knowledge of the potential health effects, while for others, such as plasticizers in food packaging materials, there is extensive toxicological data available. Yet some judgement has been made about the acceptability or otherwise of our exposure to these substances and many other substances in our environment.

The term 'Risk Assessment' has been applied to the process by which the implications for human health are evaluated. For many substances to which we may be exposed in our daily lives, some form of risk assessment, based mainly upon toxicological data, has been conducted. This is communicated in most cases by levels of acceptable intake or exposure standards. Some examples are provided in Table 1.

Many of the published standards imply a level of safety in which health risks are considered to be either 'exceedingly small' (*e.g.* UK Air Quality Standard for air pollutants), 'not appreciable' (Acceptable Daily Intake for food additives), 'not significant' [World Health Organization (WHO) drinking water guidelines] or even 'non-existent' (Occupational Exposure Standards set in the UK). While many of these standards have not been subject to a Quantitative Risk Assessment procedure, they have been set on the basis of scientific judgement and opinion, much of which is shrouded in secrecy or obscured by technical language. Over the past decade the public call for transparency and precision in the regulatory process of environmental and occupational standard setting has fuelled a need for

Table 1 *Some toxicological standards and guidelines*

Applicability	Standard	Assessment of risk
Workplace	OES	"no indication of risk to health"
	MEL	"a residual risk may exist"
Food	ADI	"without appreciable risk"
Air	AQS	"exceedingly small risk"
Soil	ICRCL	Threshold and Action Concentrations
Soil and water	EAL	"unacceptable harm" (above EAL)
	EQS	
Water	WHO Guideline	"does not result in any significant risk"

Abbreviations: OES = Occupational Exposure Standard (UK); MEL = Maximum Exposure Limit (UK); ADI = Acceptable Daily Intake; AQS = Air Quality Standard (UK); ICRCL = 'Trigger' concentrations for soil set by the Interdepartmental Committee on Reclamation of Contaminated Land (UK); EAL = Environmental Action Level (UK); EQS = Environmental Quality Standard (UK); WHO = World Health Organization.

mathematical accuracy. It is no longer accepted to state that levels are 'safe', as this now begs the response 'how safe?' Quantification of the risk assessment process offers a means of producing a satisfactory answer to such questions.

2 The Risk Assessment Process

Despite the diversity of exposure patterns and target populations to which published exposure standards are intended to apply, the risk assessment processes are broadly similar. In general, all are based on a standard *risk assessment paradigm*, such as the one published by the US National Academy of Sciences[1] in which the risk assessment process can be considered as consisting of four main steps, namely:

- *Hazard Identification*: to identify the main toxic effects (*e.g.* cancer for man)
- *Dose–Response Assessment*: to define the relationship between dose and health effects for man
- *Exposure Assessment*: to predict potential exposure levels/intake for man
- *Risk Characterization*: to define the likelihood of an adverse effect in man

These steps are described individually below.

Hazard Identification

Toxicological data from experimental animal studies is usually the primary focus of the risk assessment process, but data from epidemiological, clinical, and other sources need also to be considered, if and where available. The latter are rarely available, however, particularly when dealing with industrial chemicals, and in many cases, the hazard identification process is based solely on experimental animal studies.

The hazard identification process is concerned with defining whether the substance in question has the properties to adversely affect human health. A detailed range of animal toxicity studies would usually be required for a full characterization of toxic effects, but only in a few cases would a complete profile of effects be available. Judgements need to be taken about the weight of data and the plausibility of the effect being observed in man. Mechanistic studies are becoming increasingly important in defining whether a hazard exists for man.

For example, unleaded gasoline has been shown to cause kidney tumours in male rats. The mechanism of tumour development indicates that the presence of a specific, small, low molecular mass protein, α-2μ globulin, present in glomerular filtrate, is necessary for carcinogenicity to arise. Binding of a component found in unleaded gasoline (2,2,4-trimethylpentane) to this protein, leads ultimately to damage and destruction of the kidney tubular cells into which this chemical–protein complex is sequestered. Subsequent cell proliferation as a response to damage, leads eventually to neoplastic change and ultimately to cancer of the tubular cells. Without binding to α-2μ globulin, there is no carcinogenic response.[2] As α-2μ globulin, or any other similar protein, is not found in either female rats or humans, it is now well accepted by toxicologists that the carcinogenic effect of unleaded gasoline in male rats is not a relevant hazard for man.

A critical decision in the hazard identification process is whether the substance in question is carcinogenic or has carcinogenic potential. For purposes of risk assessments, carcinogens are divided into those that cause irreversible self-replicating DNA damage (*i.e.* genotoxic carcinogens such as benzene) and those that induce cancer through mechanisms that do not require direct interaction with DNA (*i.e.* non-genotoxic or epigenetic carcinogens such as saccharin). A genotoxic carcinogen is assumed *not* to have a threshold for initiating activity, whereas for non-genotoxic carcinogens, a threshold can be assumed to exist.

For non-carcinogens, neurotoxic or teratogenic effects are generally regarded as the most serious, as these can lead to irreversible biological damage. Adverse effects such as these are likely to be regarded as the critical or 'lead' health effect in the risk assessment process.

Dose–Response Assessment

The dose–response assessment quantitatively evaluates the toxicity information and characterizes the relationship between the dose of the substance and the incidence of health effects in the target population. The objective of this interpretation is to arrive at an exposure level for man that is 'safe', 'acceptable', or 'tolerable'. This, combined with exposure information will provide an appropriate basis for a risk assessment to be conducted.

The process usually requires predictions of the likely health effects of the substance in man from animal data. However, potential exposure levels for man are generally much lower than those used in animal studies, sometimes by several orders of magnitude. This results in two difficult areas of data interpretation over which there is much controversy:

- Extrapolation from high to low doses
- Transferability of data from animals to man

High to Low Dose Extrapolation. Toxicological studies frequently use high dosage levels, often much higher than may be expected in human exposures. Where a threshold can be demonstrated and is justifiable (as in the case for liver toxicity induced by carbon tetrachloride), it may be appropriate to assume that effects will not occur at lower doses and a No Observed Adverse Effect Level (NOAEL) can be set. From data published for carbon tetrachloride[3] a NOAEL of 1 mg kg^{-1} of body weight per day can be set, and this may be assumed to be the threshold of adverse effects in the rat.

For many substances, a threshold may not be demonstrable from available data and the only information on dose–response is at high exposure levels of the substance in question. From these data it may be possible to identify a Lowest Observed Adverse Effect Level (LOAEL), and from this predict the response or lack of it at much lower doses. A number of mathematical models and techniques have been developed for this purpose.

Where a threshold is anticipated or justified, a NOAEL may be derived from the LOAEL by taking into account the characteristics of the dose–response relationship and the mechanism of toxicity. Frequently, when mechanistic data are unavailable, a simple numerical factor may be applied to the LOAEL (say 10 or 100) to give a calculated NOAEL.

For genotoxic carcinogens, which have no threshold (see above), it is assumed that the risk equals zero only when the exposure equals zero. A number of mathematical models have been developed to extrapolate risk estimates from experimental data at high doses to give an estimate of cancer risk at each given (lower) dose level.

The models can provide linear, sublinear, or even supralinear extrapolations of experimentally derived dose–response curves to much lower levels; the choice of model depends largely upon the likely carcinogenic mechanism.[1]

The Linearized Multi-Stage (LMS) Model is the most frequently cited of all the models, and is the default option used by the Environmental Protection Agency (EPA) in the USA. This model assumes that chemical carcinogens mimic radiation in their mechanism of cancer induction, with mutation of DNA as the key event. The linearized model gives the most health-conservative estimate of dose–response, and is frequently used in risk assessment.

Transferability of Animal Data to Man. The extrapolation of dose–response characteristics from animals to man is complicated by many factors, including differences in life-span, genetic homogeneity, metabolic rates, body weight factors, metabolic pathways, bioavailability and toxic mechanisms (see Table 2). These all lead to *uncertainty* in applying data from animals to man. For example, dosage levels in toxicology studies are commonly expressed in terms of mg kg^{-1} of body weight per day. However, comparing toxicity on a body weight basis does not equate well across species. This is because, while some physiological

Table 2   *Factors to be considered when applying animal data to man*

Animal data	Human relevance
Dosage (body weight basis)	Is dose per surface area more relevant?
High doses	Same effects at low doses?
Dosage route	Relevance
Absorption	Bioavailability differences
Duration	Relevance to % of human life-span
Toxic mechanism	Applicable to man?
Genetic homogeneity	Susceptible groups
	(children, asthmatics, *etc.*)
Animal-specific effects	
Spontaneous tumours, *e.g.*	Does increased frequency indicate a risk?
mouse: liver tumours	
rat: F344 testicular tumours	
S-D mammary tumours	

parameters important in toxicological appraisal (such as liver size or creatinine clearance) are related to body weight, others (such as renal clearance, metabolic rate, activity of drug metabolizing enzymes) are not.[4] Indeed there is a better cross-species correlation when dosage is expressed in terms of body surface area.

A comparison of dosage in terms of body weight and body surface area basis is shown in Table 3. From equal bodyweight doses (mg kg^{-1}), man receives a dose 6 times higher than experienced by a rat on body surface area calculation (mg cm^{-2}) and around 9 times higher than that received by the mouse. On this basis, dividing animal body weight doses by a factor of 10 is a justifiable means of equating toxic potency from rodents to man.

Other adjustments may need to be made for life-span, duration of exposure, and route of exposure, and a number of methods for making them have been devised.[5] The question of applicability of data obtained by one exposure route to a different exposure route has been discussed by Sharratt.[6]

Furthermore, the biological significance of the observed effect in animals may not be relevant to man. Mechanistic studies have been particularly revealing in recent years for a number of substances that have been shown to induce cancer in animals, but by mechanisms either not relevant to man or by mechanisms that do not operate at low doses. Tumours caused by unleaded gasoline mentioned already arise by a mechanism that is not relevant to man; the phthalate esters (such as diethylhexyl phthalate) are not carcinogenic at low doses where there is no proliferation of peroxisomes (the increase in these intracellular organelles is necessary before cancer is induced). These effects are not considered applicable to man at low doses.

Other compounds may provoke an increase in certain types of tumours that already occur frequently in certain animal strains (e.g. testicular tumours in F344 rats, mammary tumours in Sprague Dawley rats), and generally increased carcinogenic effects in these tissues would not necessarily be considered as indicative of a cancer risk to man.

Table 3 *Comparison of dosage on body weight and surface area basis: Effects of different species receiving equal dosages of 100 mg kg⁻¹ body weight*

Species	Weight (g)	Dose (mg)	Area (cm²)	Dose/area (mg cm⁻²)	Dose/area (man:animal)
Mouse	20	2	46	0.043	9.0
Rat	200	20	325	0.061	6.4
Dog	12000	1200	5770	0.207	1.9
Man	70000	7000	18000	0.388	1.0

Similarly, metabolic pathways in animals can create active metabolites that may not be relevant to man, or that may *only* be formed at high doses, when the normal metabolic pathways are saturated. The formation of trichloroacetic acid, from 1,1,1-trichloroethane exposure, is thought to be critical to the development of liver tumours in rats, but this is not a significant pathway of metabolism in man and it throws into question the human relevance of this carcinogenic effect. While mechanistic and other information may invalidate the animal toxicity in terms of human health risk, it needs to be stressed that in the absence of such specific information, it must be assumed that effects in animals will also be seen in man.

Another factor to be considered when translating animal exposures to man is that whereas experimental animals arise from a genetically homogeneous stock, in man, the ultimate target species, there is a high degree of genetic and age-related variability in the response to toxic substances. An acceptable exposure level derived for animal data may therefore need to be modified further to take account of susceptible groups, such as children who may be more at risk from neurotoxicants, or asthmatics who may respond more severely to respiratory irritants.

All these considerations, and others, lead to *uncertainty* in appraising animal toxicity to arrive at the likely dose–response relationship in man. Any indicator of likely toxicity in man therefore needs to be qualified by consideration of what account has been taken of these factors. The careful interpretation of toxicological data can lead to a reasonable estimate of the likely dose–response relationship in man. Different approaches are taken, however, in establishing dose–response relationships for genotoxic carcinogens and those for all other substances.

Genotoxic Carcinogens. For genotoxic carcinogens, it is assumed that there is no threshold dose for initiation, the first step in a multi-stage process of carcinogenicity. Even the lowest exposure comes with a possibility that a single molecular interaction with DNA may occur and thereby initiate a genetic change that may progress to cancer, although the likelihood of this arising at very low doses may be infinitesimally small. The characteristics of the dose–response relationship in these cases allow an estimate of the level of risk to be calculated at each and every low dose level. For soil contaminants or water pollutants, where we are concerned with the effects of low level exposure, these mathematical models can be used to calculate carcinogenic risk at low doses from the response seen at high doses in animal experiments. A constraint of many of these models is that the dose–response curve is required to pass through the origin. Modelling allows the derivation of a Slope Factor (SF), a measure of how cancer risk increases (linearly) with dosage. The SF is defined by the US EPA as the 'plausible upper-bound estimate of the probability of a response per unit intake of chemical over a lifetime'.[7] The SF is used to estimate an upper-bound probability of an individual developing cancer as a result of exposure to a particular level of a potential carcinogen over a particular time period (usually a lifetime).

The SF, usually defined in terms of increased lifetime cancer risk per mg kg^{-1} day^{-1} of exposure, is calculated from available animal carcinogenicity data, most usually by the use of the LMS Model. This provides the most health conservative estimate of risk. Further conservatism is included as the value is taken from the upper 95th percentile of the confidence limit of the slope. Other factors relevant to the SF are illustrated in Table 4. It must be remembered that the risk response is linear only at low doses.

There is much controversy about the use of such models, particularly as the calculation of risk from the same data using different models can give results that vary by many orders of magnitude. For example, 11 estimates of risk from 2,3,7,8-TCDD (dioxin) conducted by US regulatory authorities in the 1980s using different models, assuming a non-threshold response and an intake of 10 pg kg^{-1} day^{-1}, varied from 1.3×10^{-3} to 7.7×10^{-18}, a difference of 10^{14} in the estimates![8] The Canadian regulatory authorities do not regard dioxin as a genotoxic carcinogen, however, and have set a threshold value of 10 pg kg^{-1} day^{-1}: this is more than 15 000 times higher than an acceptable intake of 0.006 pg kg^{-1} day^{-1} calculated by the EPA, based on an upper-bound confidence limit estimate of increased lifetime cancer risk of 1 in a million.[9,10] A reassessment of the carcinogenic risk of dioxin published by the EPA in 1994 states that the present exposure to low background levels of dioxin may be associated with even greater cancer risks, up to several hundred-fold higher than previously calculated.[11] However, the reassessment also states that the *true risk*, as opposed to the upper

Table 4   *Slope factor (SF): Characteristics and examples*

Characteristics:

- Weight of evidence (EPA groups A–E)
- Linearized Multi-stage Model as default (most health conservative)
- Based on upper 95% ile of confidence limit
- Linear at low doses only

Example values taken from IRIS database

	Slope factor (SF)	Risk model
Arsenic	15 per (mg kg^{-1} day^{-1})	Linear
Benzene	2.9×10^{-2} per (mg kg^{-1} day^{-1})	One hit

bound estimate, may actually be as low as zero (the reassessment is excellently reviewed by Conolly).[12] The dioxin risk assessment story illustrates the huge differences that can arise from different decisions about carcinogenic mechanisms of action and the appropriate extrapolation model.

Even when extrapolating from human data, large differences in risk estimates can arise. Leukaemia has been observed in industrial workers following long-term benzene exposure. From available epidemiological data, estimates[13-16] of the leukaemia risk at low levels range from around 10^{-6} to 10^{-11} per μg m^{-3}. It is important therefore to remember the variability and imprecision that are inherent in dose–risk estimates for carcinogens even when genotoxicity is not disputed.

Non-genotoxic Carcinogens and Other Substances. For substances other than genotoxic carcinogens, thresholds are assumed to exist for adverse health effects. In theory, a dose that is without health consequences for man can, therefore, be calculated. Various names have been attached to the acceptable dose for man such as Acceptable Day Intake (ADI) for food additives, or Occupational Exposure Standard (OES) for workplace chemicals. In relation to drinking water and contaminated land appraisal, the term Reference Dose (RfD) is commonly used as a measure of the toxicity threshold for environmental chemicals. The RfD is 'an estimate (with uncertainty spanning perhaps an order of magnitude or greater) of a daily exposure level for the human population, including sensitive subpopulations, that is likely to be without an appreciable risk of deleterious effects during a lifetime'.[7]

RfDs are specifically developed to be protective for long-term (chronic) exposure to a chemical, *i.e.* from 7 years to a lifetime. In addition to the Chronic RfD, the terms RfD$_s$ and RfD$_{dt}$, which refer to Subchronic and Developmental RfDs respectively, may also be used in risk assessment. In addition, RfDs can be derived for either oral or inhalation exposure. Over 600 Chronic RfDs have been reviewed and verified and have been entered onto the EPA computerized Integrated Risk Information System (IRIS).[17] The main characteristics of the RfD are illustrated in Table 5.

The RfD is derived from toxicological studies in which either the NOAEL or

Table 5   *Reference dose: Characteristics*

• Units = mg kg^{-1} day^{-1}
• Chronic (RfD), subchronic (RfD$_s$) developmental (RfD$_{dt}$)
• Oral, inhalation
• Uncertainty: can span an order of magnitude
• Similar in concept to: VSD: Virtually Safe Dose TDI: Tolerable Daily Intake MTEL: Maximum Tolerable Exposure Level ADI: Acceptable Daily Intake

Table 6   *Uncertainty factors (UFs)*

'Extrapolation'	UF	Comments
LOAEL to NOAEL	1–10	When a NOAEL cannot be derived. Value of UF depends on quality of data
Sub-chronic to Chronic	10	Used when only sub-chronic data is available
Animal to human data	10	To account for interspecies variability
Human susceptibility	10	To account for genetic and age variation in response, to protect susceptible groups (*e.g.* elderly, children)

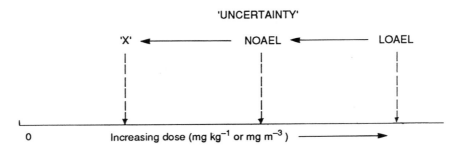

Figure 1   *Uncertainty factors: The Acceptable Human Dose "X" is derived from NOAEL or LOAEL values by the application of Uncertainty Factors. The proximity of the value "X" to zero is determined by the choice of these factors. X = Acceptable Human Dose (e.g. RfD, ADI, or OES); NOAEL = No Observed Adverse Effect Level; LOAEL = Lowest Observed Adverse Effect Level*

the LOAEL has been determined. This dose will lie between zero and the NOAEL or LOAEL, and the closeness to zero from the NOAEL or LOAEL value is determined by a number of uncertainty factors (UFs), specific to the substance in question and the quality and quantity of toxicological data available (see Figure 1). Typical UFs that may be used in defining a Chronic RfD are shown in Table 6.

Thus an RfD based on a single subchronic animal study may have a UF of between 1000 and 10 000 applied to the LOAEL. While the numbers assigned to UFs appear arbitrary there is some experimental support for these figures. For instance the ratio of subchronic to chronic NOAELs or LOAELs for 50 out of 52 chemicals has been shown to be 10 or less.[18]

In addition to the above, a modifying factor (MF) may also be applied (ranging from 1 to 10) to reflect a quantitative professional assessment of additional uncertainties in the critical study used in the entire data set that are not explicitly addressed by the UFs.

Table 7 provides some examples of RfDs obtained from NOAELs and of the UFs and MFs applied. For example, for phenol, a UF of 100 has been applied to the NOAEL of 60 mg kg^{-1} day^{-1} derived from animal studies, and comprises a

Table 7 *Basis for derivation of reference doses (RfDs) from NOAELs: Example values*

	RfD (mg kg^{-1} day^{-1})	NOAEL (mg kg^{-1} day^{-1})	UF	MF
Arsenic	0.003	0.0008	3	1
Phenol	0.6	60	100	1
Nitrite	0.1	1	1	10
Chromium (trivalent)	1	1468	100	10

Data taken from IRIS Database.

factor of 10 for interspecies extrapolation and a factor of 10 for sensitivity amongst the human population. For nitrite, no uncertainty factor was used because the NOAEL for the critical toxic effect (methaemoglobinaemia) has been derived from human data in the sensitive population (*i.e.* infants). Because nitrite is directly toxic (does not require biotransformation) a MF of 10 has been applied to the human NOAEL.

Recent information published by the Health and Safety Executive (HSE)[19,20] has also allowed an appraisal of the use of UFs for the derivation of recent UK OESs set for the protection of workers occupationally exposed to hazardous substances. The OES is theoretically equivalent to an RfD in that it represents an exposure level below which adverse effects are not expected to arise in man. As will be noted from Table 8, the UFs utilized for occupational situations are much lower than those used for RfDs, reflecting the shorter exposure times and general exclusion of susceptible groups (children, elderly) in workplace areas.[21]

Exposure Assessment

Having set a SF or RfD for a substance, Exposure Information is needed to enable Risk Characterization to take place to complete the Risk Assessment Process. A number of methodologies have been developed to derive estimates of chemical exposure from various environmental and occupational sources. The reader is referred to documentation produced by the US EPA,[7] the Welsh Development Agency,[22] and others[23] for methodologies for quantifying exposures arising from environmental contaminants. Exposure Assessment documentation

Table 8 *Uncertainty factors (UF) used by HSE in setting some Occupational Exposure Standards from animal toxicological studies*

Base of OES	UF	Example
From NOAEL ($n = 14$)	2.5–12	Chloroform (UF = 5)
From LOAEL ($n = 9$)	4–30	Cyclohexane (UF = 8)
From LOAEL ($n = 1$) (teratogenicity)	40–60	Dimethylacetamide

has also been produced by the HSE to aid in the risk assessment of newly notified substances.[24]

In terms of methodologies for the health risk assessment of contaminated sites, the exposure assessment can be regarded as comprising three main stages. Firstly, characterization of the *exposure setting*, in which the basic site characteristics and the population (present and future) who may be exposed are defined. Next, *exposure pathways* are identified, with each pathway describing the process by which a population may be exposed to chemicals at, or originating from, the site. The pathways are identified by a consideration of the sources, releases, and types and locations of chemicals at the site, *i.e.* the likely environmental fate of these chemicals (including persistence, partitioning, transport, and intermedia transfer), and the locations and activities of the potentially exposed populations. All relevant exposure routes (inhalation, ingestion, dermal contact) are identified. Finally, for each exposure pathway identified, the *magnitude, frequency, and duration of exposure* is quantified. Use is made of monitoring data and chemical transport and environmental fate models to calculate chemical intakes for the population at risk.

An end-point in the process is the calculation of a daily human intake, either as a Chronic Daily Intake (CDI) or as a Maximum Daily Intake (MDI) which is expressed in $mg \, kg^{-1} \, day^{-1}$. As upper bound estimates are frequently used in the derivation of the daily intake, this value may not be representative of what the target population is reasonably exposed to, and the uncertainties associated with these evaluations need to be fully expressed by risk assessors.

Toxicology can also make a significant contribution to the exposure assessment process as it should always be borne in mind that these crude estimates of exposure may not reflect the delivery of the effective absorbed dose to the target site. Improvements in exposure assessment to take account of pharmacodynamic, pharmacokinetic, and metabolic differences between the target population (man) and the experimental model (*e.g.* rodent) are needed to improve the validity of exposure assessment. For example, for methylene chloride, pharmacologically based pharmacokinetic modelling has shown that for the same received dose, levels of critical metabolites at the target site are up to 200 times lower for man than for animals. More detailed knowledge of toxicological mechanisms will enable improved validity of the exposure assessment process, but at present there are only a handful of chemical substances on which there is sufficient information available to enhance the exposure assessment process. Perhaps the best example is the lead exposure model which can transform environmental lead exposure data to a projected population distribution of blood lead concentrations, regarded as a direct measure of the biologically effective dose. Similarly, for exposure to formaldehyde, the measurement of DNA cross-linkages may prove to be a better measure of exposure than the inhaled dose of vapour.

Risk Characterization

The final stage in the risk assessment process is the risk characterization, in which toxicity and exposure assessments are summarized and integrated with quantitative and qualitative expressions of risk. In the risk characterization methodology utilized by the US EPA, separate assessments are made for carcinogenic effects and other chronic health effects.

For carcinogens, increased probabilities that an individual may develop cancer over a lifetime of exposure are estimated from projected intakes and the SF. For non-carcinogens, chronic health effects are characterized by comparing projected intake with the acceptable toxicity value (*e.g.* the RfD).

Tables 9 and 10 show a theoretical appraisal of health risks from water contaminated with both carcinogenic (chlordane, benzene) and non-carcinogenic (phenol, cyanide, nitrite, methyl ethyl ketone) chemicals. Major exposure pathways are via contaminated well water and the consumption of residues in fish, and the exposure assessment analysis has assigned CDIs to the various chemicals by these routes.

For the genotoxic carcinogens, benzene and chlordane, a quantified risk level is calculated by multiplying the CDI by the relevant SFs as shown in Table 9. The SF for chlordane results in an overall risk of 3×10^{-4}, arrived at by addition of the risk from the individual pathways. The acceptance, or otherwise, of this risk should include consideration that it represents the upper bound of a wide range of values, which may also include zero risk as a possibility.

For non-carcinogenic effects, some indication of whether a health risk may arise can be made by the use of the Hazard Quotient (HQ), the ratio of an exposure level to an RfD. This is a measure of how the calculated exposure compares with the acceptable toxicological level. Addition of HQs, if toxicologically plausible, to arrive at a Hazard Index (HI), can give an estimate of the acceptability of the composite exposure. HIs can be calculated separately for chronic, sub-chronic, and shorter duration exposures. In the contaminated water example shown in Table 10, individual HQs for water contaminants are calculated from the calculated CDIs and chemical-specific RfDs. Addition of the HQs gives a total HI of around 0.6. HIs of unity and above are considered to constitute a potential health risk.

There are of course other ways of appraising environmental risk by careful

Table 9 *Example of risk characterization for genotoxic carcinogens*

	CDI (mg kg^{-1} day^{-1})	SF [per (mg kg^{-1} day^{-1})]	Risk[a]
Well water			
Benzene	0.00025	0.029	7×10^{-6}
Chlordane	0.00015	1.3	2×10^{-4}
Fish			
Chlordane	0.00008	1.3	1×10^{-4}
		Total risk	$= 3 \times 10^{-4}$

[a] Calculated risk is a product of the Chronic Daily Intake (CDI) and the risk per unit of exposure given by the Slope Factor (SF). Individual risk values are added to arrive at total risk.
(Adapted from Reference 7)

Table 10 *Risk characterization: Calculation of chronic health effects*

	CDI (mg kg^{-1} day^{-1})	RfD (mg kg^{-1} day^{-1})	HQ[a]
Well water			
Phenol	0.1	0.6	0.2
Cyanide	0.0003	0.02	0.02
Nitrobenzene	0.0001	0.0005	0.2
Fish			
Phenol	0.08	0.6	0.1
Methyl ethyl ketone	0.05	0.5	0.1
	Total (Hazard Index)[b]		0.6

[a] Hazard Quotient (HQ) is the ratio of the Chronic Daily Intake (CDI) to the chronic Reference Dose (RfD) *i.e.* HQ = CDI/RfD. A value of less than unity indicates that daily intake does not exceed the threshold concentration indicated by the RfD.
(Adapted from Reference 7)
[b] Hazard Index (HI) is the sum of individual Hazard Quotients (HQ). A value of less than unity indicates the absence of chronic health risks.

examination of exposure and toxicological data, but the procedures illustrated in Tables 9 and 10 provide a visible means of quantifying the appraisal.

The acceptability of the risks, the interpretation of the risk assessment information, and communication of this information to those potentially affected are separate, detailed topics in themselves. Scientific and communication skills are necessary to ensure that information generated by the risk characterization process is correctly presented, and is not used to convey exaggerated risk.

3   Information Sources

For conducting Risk Assessments of environmental chemicals, the EPA's IRIS database is the prime source of SFs and RfDs. Peer-reviewed data on over 600 substances is contained in the database in a concise format.[17]

In the UK, IRIS can be accessed directly on-line through the British Library's BLAISE service, or alternatively hard copy printouts of IRIS Database entries can be retrieved from the British Library's Information Service at Boston Spa. Other reliable sources of information are WHO Publications (such as the Environmental Health Criteria Document for specific chemicals), HSE reviews and summaries (particularly the Criteria Document Summaries, published in the EH64 series), and the International Agency for Research on Cancer Monograph Series on the Evaluation of Carcinogenic Risks of Chemicals.

However, as the case study below illustrates, the availability of data from these sources may not be sufficient to enable health risk assessments to be carried out. Specific literature searches and evaluation of published and unpublished data by toxicologists may be necessary to derive information on acceptable levels of human exposure.

Case Study: Toxicological Information Requirements for a Quantified Risk Assessment Study of a Contaminated Land Site

In an appraisal of potential health risks to residents living on the site of a former waste disposal site, the author was involved in providing toxicological input into the quantified risk assessment study that was commissioned by the local authority. Soil analysis had revealed the presence of lead, arsenic, asbestos, and polycyclic aromatic hydrocarbons, at levels which had caused concern to the local residents.[25] Apart from children at the site, the main routes of exposure for adults were by inhalation and ingestion of airborne dust from disturbed soil, and by direct skin contact with soil. Calculation of exposure/intake from these routes, along with the use of IRIS-derived SFs and RfDs for most of the substances in question, provided a quantification of health risk. This concluded that in general the presence of contaminants in the soil did not prove a significant health risk to adults living in the vicinity, but that further detailed investigations of the risks from arsenic and lead were needed.

For children the assessment of health risk proved more problematic. Exposure modelling for this susceptible group indicated that acute soil ingestion episodes were such that if they did arise, the RfDs could be exceeded by several-fold, indicating that a potential acute ingestion hazard may exist. In the absence of published acute RfDs, tentative values were calculated from published data. In many cases these were derived from acute LD_{50} data (dose that is lethal in 50% of test subjects) and human poisoning information. From this information it was calculated that an acute toxic episode, arising from soil ingestion by a child, could arise at the site once in a calendar year.

The case study illustrates that while toxicological information for risk assessment purposes may appear to be plentiful, there are often serious data gaps when dealing with specific exposure situations that are not catered for in published information. Specialist toxicological advice is therefore a necessary part of the risk assessment process.

4   Conclusions

Toxicology has a vital and central role to play in risk assessment, not only in the obvious areas of hazard identification and definition of dose–response characteristics, but also in producing improvements both in the estimation of exposure and in the final stage of risk characterization.

Much difficulty in the risk assessment process rests on how to deal adequately with animal carcinogens. The description of SFs provided above reflects the approach taken by US regulatory authorities in attempting to assign a measure of risk to all exposure levels of a substance; from this, the increased lifetime cancer risk for populations exposed to varying concentrations of these chemicals can be calculated. In recent years, mechanistic studies of the nature of carcinogenic response to chemicals, while improving the quality of the risk assessment exercise, have also led to serious questioning of some of the default assumptions used. For instance, the assumption that the carcinogenic response in the most sensitive animal species should be used as the model for human carcinogenic response has now been questioned for a number of substances. Mice are very susceptible to the cancer-causing effects of 1,3-butadiene whereas rats are not, but while the mouse produces large amounts of the critical DNA-reactive metabolite, butadiene diepoxide, the rat and man do not[26,27]. Which species should be used as the basis for human risk assessment, the more sensitive or the more relevant?

As our understanding of toxicological mechanisms improves, we can also look forward to better estimates of exposure, ideally to provide information on levels of critical toxic metabolites at key target sites, rather than rely on gross measures of whole body exposure.

All exposures are multifactorial, and it is very rarely that exposure to only a single chemical entity takes place. When dealing with multiple exposures, it is commonplace in risk assessment methodology to use an additive formula to arrive at a composite risk estimate. This of course takes no account of potentiating, synergistic, antagonistic, or independent toxicological action and this again is another area where toxicologists need to be involved to ensure that correct judgements are made about the combined health risks from mixed chemical exposure.

Toxicologists therefore have key roles to play in improving and interpreting all of the key stages in the risk assessment process. While national bodies will have access to appropriate expertise during their risk assessment procedures, the increasingly widespread use of risk assessment methodologies at a much more local level (e.g. in assessments of contaminated land use) may be conducted without recourse to specialist toxicological advice. Plugging in numbers from the IRIS database to an exposure assessment model may produce a numerical quantitative risk estimate without specialist toxicological input. However, without an evaluation of the limitations of the data available for assessment an understanding of toxic mechanisms involved and an appreciation of the assumptions made during the risk calculations, such risk estimates may at best be misleading, and at worst they may expose the target population to an unacceptable health risk, or result in unnecessary expenditure (e.g. an unwarranted level of soil clean-up). Additionally, toxicological reference data for risk assessment purposes are limited, and are currently dictated by the availability of data on chemical compounds of environmental interest in the USA. Specialist toxicological input will be required to arrive at appropriate dose–response relationships or thresholds for substances and exposure conditions not included in the IRIS database.

Finally, having arrived at a toxicologically sound risk estimate, how do we make decisions about what action to take? The acceptability and management of risk is a topic in itself, and is beyond the scope of this chapter. The reader is, however, referred to articles on these wider implications of risk assessment by Illing,[28] Scala,[29] and Paustenbach[23] and, particularly, by Rodricks.[30]

References

1 National Research Council, 'Risk Assessment in the Federal Government. Managing the Process', National Academy Press, Washington, DC, 1983.

2 J. A. Swenberg, B. G. Short, S. J. Borgholt, J. Strasser, and M. Charbonneau, 'The comparative pathobiology of alpha (2μ)-globulin nephropathy', *Toxicol. Appl. Pharmacol.*, 1989, **97**, 35–46.

3 J. V. Bruckner, *et al.*, 'Oral toxicity of carbon tetrachloride: acute, subacute and chronic studies in rats', *Fundam. Appl. Toxicol.*, 1986, **6**, 16–34.

4 E. J. Calabrese, 'Principles of Animal Extrapolation', Wiley, New York, 1983.

5 K. S. Crump, A. Silvers, P. F. Ricci, and R. Nyzga, 'Interspecies comparison for carcinogenic potency in humans' in 'Principles of Health Risk Assessment', ed. P. Ricci, Prentice-Hall, Englewood Cliffs, NJ, 1985, ch. 8.

6 M. Sharratt, 'Assessing risks from data in other exposure routes', *Regul. Toxicol. Pharmacol.*, 1988, **8**, 399–407.

7 Environmental Protection Agency, 'Risk Assessment Guidance for Superfund', vol. 1: Human Health Evaluation Manual, US EPA, Washington, DC, 1989.

8 B. D. Beck, E. J. Calabrese, and P. D. Anderson, 'The use of toxicology in the regulatory process', in 'Principles and Methods of Toxicology', ed. A. W. Hayes, Raven Press, New York, 2nd edn, 1989, pp. 1–28.

9 Environmental Protection Agency, 'Health Effects Assessment for Polychlorinated Dibenzo-*p*-dioxin' (EPA/600/8-84/0146), US EPA, Washington DC, 1985.

10 R. J. Kociba, D. J. Keyes, J. E. Beger *et al.*, 'Results of a two-year chronic toxicity and oncogenicity study of 2,3,7,8-tetrachloridibenzo-*p*-dioxin in rats', *Toxicol. Appl. Pharmacol.*, 1978, **46**, 279–303.

11 Environmental Protection Agency, 'Health Assessment Document for 2,3,7,8-Tetrachlorobenzo-*p*-dioxin (TCDD) and Related Compounds' (EPA/600/ED-92/001a-c), US Environmental Protection Agency, Washington, DC, 1994.

12 R. B. Conolly, 'US EPA Reassessment of the Health Risks of 2,3,7,8-Tetrachlorodibenzo-*p*-dioxin (TCDD)', CIIT Activities, Chemical Industry Institute of Toxicology, Triangle Park, NC, December, 1994, pp. 1–8.

13 CONCAWE, 'Exposure and Health Risks Associated with Non-occupational Sources of Benzene', CONCAWE, Brussels, September 1994, Report No. 1/94.

14 ed. W. Sloof, 'Integrated Criterion Document – Benzene', 'National Institute for Public Health and Environmental Protection', Bilthoven, the Netherlands, 1988, Report no. 758476003.

15 M. Paxton, V. U. Chinchilli, S. M. Brett, and J. V. Rodricks, 'Leukaemia risk associated with benzene exposure in the Pliofilm cohort. II Risk estimates', *Risk Anal.*, 1994, **14**, 155–161.

16 K. Crump, 'Risk of benzene induced leukaemia: a sensitivity analysis of the Pliofilm cohort with additional follow-up and new exposure estimates', *Toxicol. Environ. Health*, 1994, **42**, 219–242.

17 Environmental Protection Agency, Integrated Risk Information System, On-line Database, US EPA, Washington, DC.

18 M. L. Dourson and J. F. Stara, 'Regulatory history and experimental support of uncertainty (safety) factors', *Regul. Toxicol. Pharmacol.*, 1983, **3**, 224–238.

19 Health and Safety Executive, 'Criteria Document Summaries. Synopses of the Data Used in Setting Occupational Exposure Limits. 1993 Edition', Guidance Note EH64, HSE, 1993.

20 Health and Safety Executive, 'Criteria Document Summaries. Synopses of the Data Used in Setting Occupational Exposure Limits. 1994 Supplement', Guidance Note EH64, 1995 Supplement, HSE, 1995.

21 S. Fairhurst, 'The uncertainty factor in the setting of Occupational Exposure Standards', *Ann. Occup. Hyg.*, 1995, **39**, 375–385.

22 Welsh Development Agency, 'The WDA Manual on the Remediation of Contaminated Land', (Welsh Development Agency/ECOTOC Research & Consulting Ltd/Environmental Advisory Unit Ltd), Welsh Development Agency, Cardiff, 1994.

23 D. J. Paustenbach, 'A survey of health risk assessment', in 'The Risk Assessment of Environmental and Human Health Hazards: A Textbook of Case Studies', ed. D. J. Paustenbach, John Wiley, 1994, pp. 27–124.

24 Health and Safety Executive, 'Risk Assessment of Notified New Substances. Technical Guidance Document', including EASE software (Estimation and Assessment of Substance Exposure), HSE Books, London, 1994.

25 M. Quint and A. Limage, 'Risk Assessment Methodology at Glory Hole, Portsmouth', in IWEM Proceedings of Conference 'Contaminated Land: from Liability to Asset', Birmingham, 1994,

26 M. W. Himmelstein, M. J. Turner, B. Asgharian, and J. A. Bond, 'Comparison of blood concentrations of 1,3-butadiene and butadiene epoxides in mice and rats exposed to 1,3-butadiene by inhalation', *Carcinogenesis*, 1994, **15**, 1479–1486.

27 M. A. Medinsky, T. L. Leavens, G. A. Csanady, M. L. Gargas, and J. A. Bond, 'In vivo metabolism of butadiene by mice and rats: a comparison of physiological model predictions and experimental data', *Carcinogenesis*, 1994, **15**, 1329–1340.

28 H. P. A. Illing, 'Possible risk considerations for toxic risk assessment', *Hum. Exp. Toxicol.*, 1991, **10**, 215–219.

29 R. A. Scala, 'Risk assessment', in 'Cassarett and Doull's Toxicology', 4th edn, ed. M. O. Amdur, J. Doull, and C. D. Klaasen, Pergamon Press, New York, 1989, pp. 984–996.

30 J. V. Rodricks, 'Calculated Risks: The Toxicity and Human Health Risks of Chemicals in our Environment', Cambridge University Press, Cambridge, UK, 1992.

Chapter 3

Toxicological Information from Animal Data: Methylene Chloride

By Trevor Green

ZENECA CENTRAL TOXICOLOGY LABORATORY, ALDERLEY PARK,
MACCLESFIELD, CHESHIRE, SK10 4TJ, UK

1   Introduction

In 1986 the US National Toxicology Program[1] reported the results of a 2 year cancer bioassay in which B6C3F1 mice and F344 rats had been exposed to methylene chloride (dichloromethane) by inhalation at concentrations of 2000 and 4000 ppm. Significantly elevated incidences of lung and liver tumours were seen in exposed mice, but not rats, and in a previous study the incidences of these tumours were not increased in rats or hamsters exposed under similar conditions.[2] Methylene chloride, in addition to its use as an industrial cleaning agent and process solvent, is also sold into the consumer market in the form of paint stripper, as a solvent in aerosols, and as an indirect food additive through its use in decaffeination of coffee and in a number of other food processes. Consequently, the increased tumour incidences seen in mice gave cause for concern about the risks to human health, although the lack of effect in rats and hamsters also gave rise to considerable uncertainty as to the true hazard to humans exposed to this chemical.

In response to this uncertainty, a number of industries producing methylene chloride sponsored, through their respective trade associations (the European Chlorinated Solvent Association, the Halogenated Solvents Industry Alliance, and the Japan Association for Hygiene of Chlorinated Solvents), a series of studies to explain the basis of the species differences in laboratory animals and the relevance of the results in mice to humans.

2   Genotoxicity and Related Assays

An assessment of the genotoxic potential of a chemical carcinogen is a key part of any mechanistic study and human hazard assessment. Methylene chloride has been tested for genotoxicity in a wide range of gene mutation and chromosomal assays in prokaryotic and eukaryotic systems *in vitro* and *in vivo* (for a review see reference 3).

Methylene chloride is clearly mutagenic in bacteria and weakly positive responses have been seen in some strains of yeast, a eukaryotic micro-organism. The results of gene mutation and UDS assays in mammalian cells and *in vivo* were uniformly negative. Chromosomal aberrations have been reported following exposure of a variety of cell types *in vitro* to very high concentrations of methylene chloride. *In vivo*, there was no evidence of chromosomal damage in rat or mouse bone marrow, or in *Drosophila*.[3] In a more recent report, small increases in chromosomal damage were seen in lung cells and circulating lymphocytes of mice exposed to 8000 ppm methylene chloride.[4]

Studies isolating DNA from target tissues of rats and mice exposed to radiolabelled methylene chloride failed to find evidence of DNA binding in the livers or lungs of mice, rats, or hamsters.[5,6] No convincing evidence of mitogenesis was found in the livers of treated mice,[7,8] nor was there any definitive evidence of oncogene activation in either liver or lung tissue taken from mice during a 2 year study conducted by the US National Institute for Environmental Health Sciences (NIEHS).[9]

As a guide to the mechanism of action of methylene chloride as a mouse carcinogen, the results of this considerable range of tests for genotoxicity and chromosomal damage were inconclusive. Methylene chloride clearly has the potential to interact with DNA, as exemplified by the bacterial mutagenicity, yet there was no evidence to suggest that this potential was expressed in mammalian cells either *in vitro* or *in vivo*. However, the target organs in the mouse are not normally used, either *in vitro* or *in vivo*, for any of the standard gene mutation assays and the potential for methylene chloride or its metabolites to interact with DNA in these tissues was unknown.

3   Target Organ Toxicity

A number of mechanisms of carcinogenesis are now known that do not involve a direct interaction between a chemical or its metabolites and DNA. Cell damage, particularly in conjunction with cell division (S-phase) may result in an increased expression of the high background tumour rates found in the liver and lungs of B6C3F1 mice. Following exposure to 2000 and 4000 ppm methylene chloride, 6 h day^{-1} for 10 days, there was evidence of a dose-dependent increase in liver/bodyweight ratios (Hext *et al.*, 1986, unpublished work). During a 90 day study using a standard bioassay exposure protocol (6 h day^{-1}; 5 days per week), highly selective damage to the mouse lung bronchiolar Clara cell was seen after a single exposure, which recovered on repeated exposure.[10] The damage seen after a single exposure was repeated after the first exposure of each week and was accompanied during the first 2 weeks of the study by an increase in the number of cells in S-phase. The liver growth and lung damage seen in mice were not seen in rats exposed to the same dose levels.

These studies failed to provide any clear indication of a mechanism of action in the mouse other than evidence of liver growth and pulmonary damage accompanied by increased cell division. Although these effects could potentially result in an increased expression of background tumour rates, or 'promotion' of other underlying mechanisms, their exact contribution to the development of the mouse tumours is not known. The responses were, however, clearly specific to the mouse and were not seen in rats at equivalent dose levels.

4 Metabolism and Pharmacokinetics

Methylene chloride is metabolized by two pathways[11-15], one catalysed by cytochrome P-450 enzymes, the other by glutathione S-transferases. The oxidative cytochrome P-450 pathway produces carbon monoxide and carbon dioxide via an unstable intermediary metabolite, formyl chloride; the glutathione pathway yields carbon dioxide following the formation of a postulated glutathione conjugate and formaldehyde (Figure 1). Neither formyl chloride nor the glutathione conjugate of methylene chloride have been isolated or characterized, although their formation is entirely consistent with the product formed and the enzymes and cofactors required. *In vitro*, the two pathways can be conveniently separated by tissue fractionation techniques. Assays of metabolic activity *in vitro* are based on carbon monoxide formation from the cytochromes P-450 pathway and formaldehyde formation from the glutathione S-transferase route. There is now evidence, that was not available when these studies were conducted, that the isoenzymes responsible for methylene chloride metabolism by these pathways are principally cytochrome P-450IIE1[16] and the glutathione S-transferase theta class enzymes.[17]

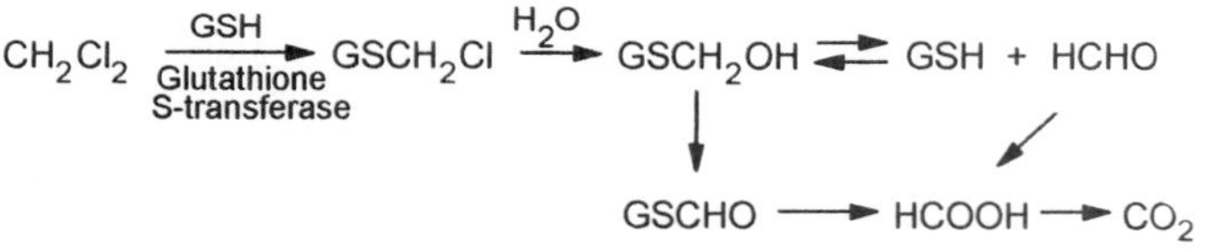

Figure 1   *Methylene chloride metabolic pathways (GSH = reduced glutathione)*

The metabolism of methylene chloride by the two pathways was determined in B6C3F1 mice and F344 rats during and following 6 h exposures to concentrations ranging from 100 to 4000 ppm.[18] From these studies, the utilization of the two pathways at different dose levels was determined (Figure 2) and a comparison of the rates of metabolism in the two species obtained.

In summary these studies provided evidence for the following:

(a) Circulating levels of methylene chloride in blood were 5-fold higher in rats

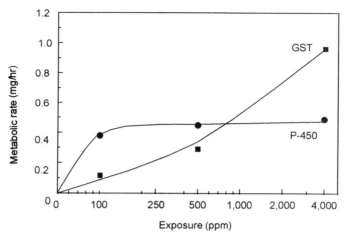

Figure 2   *The dose-dependent metabolism of methylene chloride in the mouse*
          (GST = glutathione S-transferase)

than mice at the 2000 and 4000 ppm dose levels used in the National
Toxicology Program[1] study.

(b) The cytochrome P-450 pathway was the major route of metabolism in both
species at low dose levels; it became saturated above doses between 100 and
500 ppm and was quantitatively similar in rats and mice.

(c) The glutathione S-transferase pathway was a major pathway only in mice,
its activity at the 4000 ppm dose level being more than an order of
magnitude greater than in rats.

(d) The route of metabolism is markedly dose-dependent.

These studies pointed to the glutathione pathway as the major species
difference between mice and rats which could be related to carcinogenicity.
Utilization of the cytochrome P-450 pathway was similar in both species and
circulating levels of parent chemical were considerably higher in the non-target
species, the rat, than in mice.

Following these *in vivo* studies, further comparisons of the two metabolic
pathways were made using liver fractions from mice, rats, hamsters, and
humans.[18] The rat–mouse comparisons reflected the differences in the two
pathways seen *in vivo* indicating that measurements made *in vitro* in hamster and
human tissues, where *in vivo* data was not available, would similarly be
representative. These studies measuring *in vitro* metabolic rates (K_m and V_{max})
revealed similar rates for all four species for the cytochrome P-450 pathway, but
markedly different rates for the glutathione S-transferase pathway (Figure 3).

Subsequently, Reitz *et al.*,[19] Bogaards *et al.*,[20] and Graves *et al.*[21] reported the
results of studies in which further human liver samples were assayed for
glutathione S-transferase activity with methylene chloride. A summary of all of
the currently available human liver assays is shown in Figure 4.

From this detailed comparison of the metabolism and pharmacokinetics of

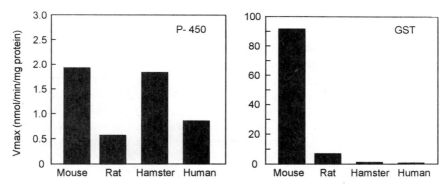

Figure 3 *The in* vitro *metabolism of methylene chloride in liver fractions from mice, rats, hamsters, and humans (GST = glutathione S-transferase)*

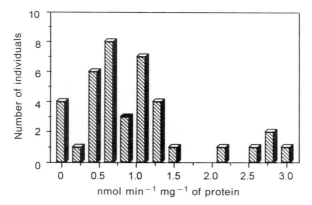

Figure 4 *The metabolism of methylene chloride by the glutathione S-transferase pathway in 39 human liver samples. Data taken from references 19–21*

methylene chloride it was concluded that the species-specific liver and lung tumours seen in mice resulted from the very high level of glutathione *S*-transferase metabolism of methylene chloride, which is unique to this species.

5 Physiologically Based Pharmacokinetic (PB-PK) Modelling and Risk Assessment

The species differences and complex dose-dependent behaviour of the two metabolic pathways revealed by the above studies are not described by the default assumptions conventionally used in quantitative risk assessment. Alternative procedures which could incorporate the pharmacokinetic data described above were considered to give a more accurate assessment of human risk.

To that end, a PB-PK model was constructed using the above data to describe the species-dependent metabolism of methylene chloride over a wide range of dose levels.[22] The ability of the model to predict the *in vivo* behaviour of methylene chloride was validated against *in vivo* animal studies. It was also validated against a number of human volunteer studies reported in the literature, in which volunteers were exposed to concentrations of methylene chloride from 50 to 750 ppm, for periods up to 7.5 h, using complex protocols and varying work rates. Parameters measured included pulmonary uptake, venous blood concentration, and post-exposure elimination of methylene chloride and carbon monoxide. The fact that the model was able to provide an accurate description of this diverse data set provided a high degree of confidence that such a model was a sound basis for risk assessment.

Using the PB-PK model, risks were based on the concentrations of glutathione *S*-transferase metabolites in the target organs of the species of interest as a more meaningful alternative to the external dose of methylene chloride. This process resulted in a steeper dose–response curve and a reduction in the risks estimated to occur at low dose levels by several orders of magnitude, when compared with the Global 82 model using linearized extrapolation and external dose.[22]

6   Mechanism of Action

The excellent correlation between glutathione *S*-transferase metabolism of methylene chloride and the occurrence of tumours in animals provided an explanation for the observed species differences, but did not explain the mechanism of action in the mouse. The roles of genotoxicity and cytotoxicity were at best ambiguous and there was no clear indication of what the mechanism of action might be. A series of studies conducted by the NIEHS looking at oncogenes and other markers in methylene chloride-induced tumour tissue failed to clarify definitively the relative roles of genotoxic and non-genotoxic mechanisms.[8,9,23]

Genotoxicity *In Vitro*

The results of the bacterial mutation assays suggested that methylene chloride has the intrinsic property of genotoxicity. The glutathione *S*-transferase pathway produces two potentially reactive intermediates, the glutathione conjugate *S*-chloromethyl glutathione, and formaldehyde, a known mutagen. The glutathione conjugate, which is structurally similar to the reactive and mutagenic conjugates of 1,2-dihaloethanes, is extremely unstable, has never been isolated, and its reactivity towards biological molecules never established. Rapid hydrolysis and decomposition leads to the formation of formaldehyde (and glutathione; Figure 1).

Methylene chloride is mutagenic in *Salmonella typhimurium* TA100 and in *Escherichia coli*, and both systems have been used to identify the metabolites responsible for this mutagenicity.[24] Bacteria had previously been shown to be capable of metabolizing methylene chloride[25] and were subsequently found to contain an enzyme of the glutathione *S*-transferase family.[26] Glutathione deficient strains of *S. typhimurium* TA100 were also used to show that methylene chloride mutagenicity was glutathione-dependent, thus further linking the response to the glutathione pathway.[24] Thier *et al.*[27] provided final confirmation by expressing a rat glutathione *S*-transferase enzyme in *Salmonella* and demonstrating a marked increase in methylene chloride mutagenicity. The lack of sensitivity of *Salmonella* TA100 towards formaldehyde led to the conclusion that the glutathione conjugate was responsible for the mutagenicity in this system. However, a comparison of the sensitivity of DNA-repair-deficient mutants of *E. coli* towards methylene chloride, formaldehyde, and 1,2-dibromoethane revealed that the mutagenic responses seen with methylene chloride could also be attributed to its metabolism for formaldehyde.[24]

DNA damage in the form of single strand breaks and DNA–protein cross-links were detected in Chinese hamster ovary (CHO) cells incubated in the presence of methylene chloride and mouse liver fractions.[28] These responses were linked to the glutathione pathway by comparing the ability of mouse liver microsomal and cytosolic fractions to activate methylene chloride. Only the cytosol fraction was able to metabolize methylene chloride to DNA damaging species. Semicarbazide, a formaldehyde-trapping reagent, was used to distinguish between effects caused by formaldehyde and those caused by the glutathione conjugate of methylene chloride. DNA–protein cross-links were abolished by the addition of semicarbazide, but not DNA single strand breaks, showing that formaldehyde was responsible for the cross-linking but not the strand breaks. The latter were attributed to the glutathione conjugate.[28]

When freshly isolated hepatocytes from mice and rats were incubated with methylene chloride, DNA single strand breaks were detectable by alkaline elution analysis.[28] Although breaks were detected in cells from both species, there was a marked difference in sensitivity between mouse and rat hepatocytes to methylene chloride. In the mouse, DNA damage could be detected at concentrations as low as 0.4 mM, whereas 30 mM concentrations had to be used to induce the same response in rat hepatocytes. Hepatocytes that had been depleted of glutathione were less sensitive to the DNA damaging effects of methylene chloride. Formaldehyde was also able to induce DNA single strand breaks in hepatocytes. However, the concentration required would not have been achieved by the metabolism of methylene chloride, leading to the conclusion that these effects were caused by the glutathione conjugate and not by formaldehyde.[28]

Both mouse and rat hepatocytes responded to methylene chloride but with markedly different sensitivities. Conversion of the *in vitro* concentrations used in these assays to *in vivo* equivalent exposures by means of the methylene chloride PB-PK model found the concentrations used with the mouse cells to be equivalent to 4000 ppm and those with the rat to 85 000 ppm. Thus, the results were consistent with the outcome of the cancer bioassays. DNA damage was seen in mouse hepatocytes at dose levels equivalent to those used in the bioassays, whereas the dose equivalent in the rat was several fold higher than a lethal dose ($LD_{50} \approx 15\,000$ ppm) and could not be achieved *in vivo*.

Of the 40 or so lung cell types the mouse lung Clara cell is the only one known to be affected by exposure to methylene chloride. Clara cells were therefore chosen for a series of experiments equivalent to those completed in hepatocytes.

Methylene chloride was found to cause a dose-dependent increase in DNA single strand breaks in Clara cells following *in vitro* exposure. Glutathione depletion experiments linked the strand breaks to the glutathione *S*-transferase pathway.[21] Thus, identical responses were seen in cells from the target organs in the mouse suggesting a common mechanism for the development of both liver and lung tumours.

Genotoxicity *In Vivo*

The DNA damaging effects of methylene chloride seen in mouse liver and lung cells *in vitro* were reproduced *in vivo* following exposure of mice to methylene chloride at the concentrations used in the cancer bioassays.[21] DNA single strand breaks were detected in hepatocytes and whole lung homogenates prepared from mice following 6 h exposures. Pre-treatment of mice with a glutathione-depleting agent (buthionine sulfoximine) reduced the level of DNA damage supporting the link between this damage and the glutathione pathway established in the previous studies. As predicted from the *in vitro* experiments, DNA damage could not be detected in either the lungs or the livers of rats exposed *in vivo* to methylene chloride.

Enzyme Studies

The metabolism and genotoxicity studies have clearly linked the mutagenicity and carcinogenicity of methylene chloride in the mouse to metabolites of the glutathione *S*-transferase pathway. The rate of metabolism by this pathway is also linked to the species differences in carcinogenicity. Consequently, the nature, activity, and localization of the enzyme(s) that metabolize methylene chloride via this pathway are a key feature in human safety assessment. Two isoenzymes, equivalent to those of rat theta class transferases 5-5 and 12-12, have been isolated from mouse liver.[29] Glutathione *S*-transferase 5-5 was found to be the major enzyme involved in the metabolism of methylene chloride. Both enzymes have been sequenced and their distribution is being studied in liver and lung samples from mice, rats, and humans using complementary DNA probes and antibodies.

Conclusions Concerning the Mechanism of Action

Methylene chloride causes liver and lung tumours in mice as a result of interactions between metabolites of the glutathione *S*-transferase pathway and DNA. These interactions have been characterized in a range of systems from micro-organisms *in vitro* to mice *in vivo*. Alkylation of DNA by the glutathione conjugate, *S*-chloromethyl glutathione, leads to mutations that have been characterized in CHO cells, and it is reasonable to assume that this DNA alkylation also leads to the strand breaks seen in these cells and in mice *in vivo*. Although DNA–protein cross-links have not been measured *in vivo* in these studies, they have been detected by others,[30] and a possible role for cross-linking cannot be ruled out. However, it seems probable that the critical event *in vivo* is that between the glutathione conjugate and DNA, which leads to DNA mutation. In addition to DNA damage, methylene chloride induced liver growth, Clara cell damage, and increases in S-phase in the bronchiolar epithelium are factors which may further increase the susceptibility of mice to developing tumours. None of the above effects can be detected in rats *in vivo*.

7 Species Differences

A plausible mechanism of action for methylene chloride as a mutagen and carcinogen has been established in the mouse. Part of that process involved comparisons with the rat and clear differences were established for the end-points measured. This work was then extended to include the hamster, the third species used in the cancer studies, and finally human tissues.[21] DNA single strand breaks could not be detected in either hamster or human hepatocytes ($n=8$) using maximal (cytotoxic) concentrations of methylene chloride. The highest concentrations used with human hepatocytes were determined by PB-PK modelling to be equivalent to whole body exposures of 250 000 ppm, a dose that is two orders of magnitude higher than a lethal dose. A comparison of mice, rats, and humans is shown in Figure 5. The ability of the human hepatocytes used in these studies to

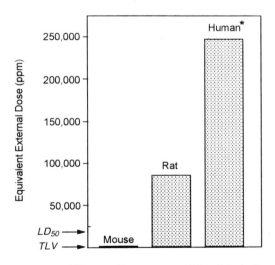

* Human: No response at this cytotoxic dose

Figure 5   *DNA single strand breaks in hepatocytes incubated with methylene chloride. Hepatocytes were incubated with methylene chloride in vitro and DNA strand breaks measured by alkaline elution. The concentration of methylene chloride used in these experiments was converted to an equivalent inhalation dose using the PB-PK model described in Section 5 (LD_{50} = dose that is lethal in 50% of subjects; TLV = threshold limit value)*

metabolize methylene chloride was typical of that seen with a wide range of human liver samples[21] (Figure 4). The same cells were extremely sensitive to 1,2-dibromoethane, strand breaks being detected at concentrations as low as 0.5 mM.

8   Human Hazard and Risk Assessment

The animal cancer data and the human epidemiology studies show that the lung and liver cancer seen in mice is unique to that species. This research programme has provided clear explanations for why these tumours are restricted to the mouse and for the mechanisms involved. They may be summarized as follows.

(a) Methylene chloride is metabolized by two metabolic pathways. At low dose levels it is metabolized mainly by cytochrome P-450 at rates that do not differ markedly between species. The glutathione S-transferase pathway is a major pathway only in mice and then only at high dose levels. A more detailed analysis of the distribution of the transferase enzymes in liver and lung tissues from mice, rats, and humans is currently being carried out.

(b) The tumours in mice are caused by a genotoxic mechanism involving metabolites of the glutathione S-transferase pathway, principally the glutathione conjugate S-chloromethylglutathione. The mechanism is common to both liver and lung.

(c) S-Chloromethylglutathione is reactive but highly unstable. The ability of this unstable intermediate to act as a DNA alkylating agent is highly dependent upon its rate and site of formation. High enzyme activity within certain cells and within nuclei account for the markedly higher sensitivity of the mouse to the genotoxic effects of methylene chloride. In the absence of high or localized glutathione S-transferase activity, the metabolites of the glutathione S-transferase pathway have no measurable mutagenicity, for example in short-term mammalian mutagenicity tests, in the rat *in vivo*, and in rat, hamster, and human tissues *in vitro*.

(d) Liver growth, cellular damage and increases in cell division in the lung are additional risk factors seen only in the mouse.

From this overview of the toxicology of methylene chloride it is clear that the mouse is atypical of other species, including humans. The entire data base, including the mutagenicity tests, cancer bioassays, human epidemiology studies, metabolism studies *in vivo* and *in vitro*, the activity of the transferase enzymes, and the DNA damaging effects of the metabolites of the glutathione pathway, is consistent with the mechanism and species differences described. The B6C3F1 mouse has been found to be unique in its response to methylene chloride, and it cannot therefore be considered to be an appropriate model for humans. Hence, it is inappropriate to use mouse liver and lung tumours as the basis for methylene chloride hazard and risk assessments.

References

1 NTP, National Toxicology Program TR 306, Final Report, NIH Publication No. 86–2562, Washington, DC, 1986.
2 J. D. Burek, K. D. Nitschke, T. J. Bell, D. L. Wackerle, R. C. Childs, J. E. Beyer, D. A. Dittenber, L. W. Rampy, and M. J. McKenna, *Fundam. Appl. Toxicol.*, 1984, **4**, 30.
3 ECETOC, Technical Report No. 26, ECETOC, Brussels, 1987.
4 J. Allen, A. Kligerman, J. Campbell, B. Westbrook-Collins, G. Erexson, F. Kari, and E. Zeiger, *Environ. Mol. Mutagen.*, 1990, **15**, 221.
5 A. M. Schumann, T. R. Fox, K. D. Nitschke, and P. G. Watanabe, *The Toxicologist*, 1984, **4**, 445.
6 T. Green, W. M. Provan, D. C. Collinge, and A. E. Guest, *Toxicol. Appl. Pharmacol.*, 1988, **93**, 1.
7 P. A. Lefevre and J. Ashby, *Carcinogenesis*, 1989, **10**, 1067.
8 J. F. Foley, P. D. Tuck, T.-V. T. Ton, M. Frost, F. Kari, M. W. Anderson, and R. R. Maronpot, *Carcinogenesis*, 1993, **14**, 811.
9 T. R. Devereux, J. F. Foley, R. R. Maronpot, F. Kari, and M. Anderson, *Carcinogenesis*, 1993, **14**, 795.
10 J. R. Foster, T. Green, L. L. Smith, R. W. Lewis, P. M. Hext, and I. Wyatt, *Fundam. Appl. Toxicol.*, 1992, **18**, 376.
11 V. L. Kubic and M. W. Anders, *Drug Metab. Dispos.*, 1975, **3**, 104.
12 V. L. Kubic and M. W. Anders, *Biochem. Pharmacol.*, 1978, **27**, 2349.
13 A. E. Ahmed and M. W. Anders, *Drug Metab. Dispos.*, 1976, **4**, 357.
14 A. E. Ahmed and M. W. Anders, *Biochem. Pharmacol.*, 1978, **27**, 2021.
15 M. L. Gargas, H. J. Clewell, and M. E. Andersen, *Toxicol. Appl. Pharmacol.*, 1986, **82**, 211.
16 F. P. Guengerich, D. H. Kim, and M. Iwasai, *Chem. Res. Toxicol.*, 1991, **4**, 168.
17 D. J. Meyer, B. Coles, S. Pemble, K. S. Gilmore, G. M. Fraser, and B. Ketterer, *Biochem. J.*, 1991, **274**, 409–414,
18 T. Green, 'Biologically-Based Methods for Cancer Risk Assessment', ed. C. C. Travis, Plenum, New York, 1989, p. 289.
19 R. H. Reitz, A. L. Mendrala, and F. P. Guengerich, *Toxicol. Appl. Pharmacol.*, 1989, **97**, 230.
20 J. J. P. Bogaards, B. van Ommen, and P. J. van Bladeren, *Biochem. Pharmacol.*, 1993, **45**, 2166.
21 R. J. Graves, C. Coutts, and T. Green, *Carcinogenesis*, 1995, **16**, 1919.
22 ECETOC, Technical Report No. 32, ECETOC, Brussels, 1988.
23 F. W. Kari, J. F. Foley, S. K. Seilkop, R. R. Maronpot, and M. W. Anderson, *Carcinogenesis*, 1993, **14**, 819.
24 R. J. Graves, R. D. Callander, and T. Green, *Mutat. Res.*, 1994, **320**, 235.
25 T. Green, *Mutat. Res.*, 1983, **118**, 277.
26 S. D. La Roche and T. Leisinger, *J. Bacteriol.*, 1990, **172**, 164.
27 R. Thier, J. B. Taylor, S. E. Pemble, W. G. Humphreys, M. Persmark, B. Ketterer, and F. P. Guengerich, *Proc. Natl. Acad. Sci. U.S.A.*, 1993, **90**, 8576.
28 R. J. Graves, C. Coutts, H. Eyton-Jones, and T. Green, *Carcinogenesis*, 1994, **15**, 991.
29 G. W. Mainwaring, M. Davidson, and T. Green, *Biochem. J.*, 1996, in press.
30 M. Casanova, D. F. Deyo, and H. Heck, *Toxicol. Appl. Pharmacol.*, 1992, **14**, 162.

Chapter 4

Dealing with Genotoxic Carcinogens: A UK Approach

By Anne McDonald*

HEALTH ASPECTS OF ENVIRONMENT AND FOOD DIVISION, DEPARTMENT OF HEALTH, SKIPTON HOUSE, 80 LONDON ROAD, ELEPHANT AND CASTLE, LONDON SE1 6LW, UK

1 Introduction

Numerous substances are known to be capable of causing cancer in experimental animals and several have been shown to do so in humans. The International Agency for Research on Cancer (IARC) recognizes 58 agents (the term includes chemical compounds, physical agents, such as radiation, and biological factors, such as viruses), groups of agents, mixtures of agents, or 'exposure circumstances' (related to occupation or personal and cultural habits, *e.g.* smoking) as being carcinogenic to humans. Many other agents considered by IARC working groups are judged as probable or possible human carcinogens, based on the strength of the evidence from studies in humans and experimental animals and other relevant data.[1] These other data often include short-term mutagenicity tests for genotoxicity. The terms genotoxic carcinogen and non-genotoxic carcinogen are very widely used, although the distinction may not be straightforward. In this chapter genotoxic carcinogens are defined as those that cause DNA damage, either directly or after metabolic activation. Non-genotoxic carcinogens as a group are difficult to define, apart from negatively, *i.e.* chemicals that induce cancer but show no activity in a series of well-conducted, fully validated genotoxicity tests.

Non-genotoxic carcinogens act by a diverse range of mechanisms, some of which are poorly understood. For regulatory purposes in the UK a precautionary approach is taken. Animal carcinogens that do not appear to be genotoxic in short-term mutagenicity tests, but for which no mechanism (for which a threshold could be defined) of carcinogenic action has been established, are treated in the same way as genotoxic carcinogens. If a compound is to be regarded as a non-genotoxic carcinogen the evidence for the proposed mechanism of non-genotoxic carcinogenesis must first be assessed. If the mechanism of carcinogenesis

---

* The opinions expressed in this paper are those of the author and should not be taken as representing those of the Department of Health.

is well understood, then consideration needs to be given to whether the mechanism can be: (1) extrapolated from high to low doses in the species in which the effects are seen; and (2) extrapolated between species (*i.e.* is the mechanism relevant to humans).

Many non-genotoxic carcinogens act through a primary toxic effect, which, in the best understood examples, induces repeated cycles of cell damage and regeneration. Thus for these substances there is no carcinogenic hazard at doses that do not produce the primary toxic event, and a threshold can be set for the induction of carcinogenesis. In other cases the mechanism of carcinogenicity, which induces tumours in the laboratory animal, is not relevant in humans; *e.g.* in the male rat kidney malignancies are induced by compounds such as 1,4-dichlorobenzene interacting with α-2u globulin, a protein essentially specific to male rats. Regulations based on such an effect are not needed.

2   Genotoxic Carcinogens

Regulatory authorities within the UK accept the prudent assumption that genotoxic carcinogens have no threshold of effect, *i.e.* any exposure will result in some increase in risk. This is based on current understanding of the mechanism of action of genotoxic carcinogens, *i.e.* that the damage to DNA caused by the genotoxic agent (or metabolite) will lead to a mutation in a gene that is important in the carcinogenic process, leading to the development of a malignant tumour. Theoretically one molecule of the genotoxic compound could lead to such a mutation and thus ultimately to a tumour.

There are in fact a number of defence mechanisms and self-limiting processes within the cell that prevent the transmission of mutated DNA to daughter cells. For example, damaged DNA may be repaired efficiently by error-free DNA repair, or the damage may be so extensive that the cell cannot replicate and dies. These and other defence mechanisms that prevent the development of malignancies mean that there probably is a threshold of effect for genotoxic carcinogens, but detection of this threshold in animal studies is not possible because of the difficulty of demonstrating the dose–response relationship at very low doses.[2,3] It is also impossible to prove the converse, *i.e.* that any given case of cancer has *not* arisen as a result of exposure to a single molecule. This, together with the lack of experimental means of defining a threshold, has led to the adoption of the 'no safe level' approach for genotoxic carcinogens.

Risk estimates for genotoxic carcinogens generally rely upon extrapolation from high experimental doses in laboratory rodents to low exposures in humans. Relatively high doses are necessary in animal experiments because of the low sensitivity of conventional experiments for the detection of carcinogens. However, even very large experiments in terms of numbers of animals are insufficiently sensitive to define the dose–response relationship at the lower end of the curve. Long-term carcinogenicity tests are conducted in well-defined strains of rats and mice, so that there is detailed historical knowledge of the background incidence of spontaneous tumours. The sensitivity of the bioassay to carcinogens is even less in tissues with high spontaneous tumour rates. The poor sensitivity in these studies is compensated for by the use of high doses, up to the maximum tolerated dose, leading to further problems when extrapolating from higher to lower doses, since there may be distortion or overloading of metabolic pathways at the higher doses.

One method that has been widely used, particularly in the USA, to extrapolate from experimental doses in laboratory rodents to the low exposures experienced by humans is the fitting of mathematical models to the dose–response curve for the animal bioassay and using the model to predict the response at low doses. In regulatory toxicological practice in the UK such mathematical quantitative risk assessment (QRA) methods are not routinely used. In general, UK regulatory authorities consider all the available data concerning the carcinogenicity of the substance, *i.e.* chemical structure, metabolic and distribution data, genotoxicity, carcinogenicity in lifetime animal studies, and epidemiological evidence derived from human exposures. Where these data include the results of QRA calculations, those results may be considered along with the other information. If the 'weight of evidence' leads to the conclusion that the compound is a potential human carcinogen with a genotoxic mechanism, then the overall aim is to seek to eliminate exposure or, if this is not possible, to reduce exposure as far as is possible. It is possible that the benefits of the use of a compound are such that the benefits outweigh the carcinogenic risks, *e.g.* the use of therapeutic alkylating agents in the treatment of cancer, but exposure should be reduced as far as possible in those not directly benefitting from the use (medical staff in this example). A substance that is an *in vivo* mutagen in animals will be assumed to be a potential genotoxic carcinogen and handled in the same way, unless there is very good evidence that this will not be the case in humans.

This approach is pragmatic, and the measures taken to reduce risk depend on the compound involved. The action taken is based on the ease with which exposure can be reduced, the cost of the reduction, and the benefit from use of the compound. Although this could lead to a charge of inconsistency, this approach does avoid the use of numerical risk estimates of doubtful accuracy. The UK Committee on Carcinogenicity of Chemicals in Food, Consumer Products, and the Environment reviewed the use of mathematical models for risk assessment in 1991.[4] The Committee did not endorse their use because all the models were based on mathematical assumptions rather than a knowledge of biological mechanisms and, depending on the model used, a wide range of low-dose risk estimates could be generated from the same experimental data. None of the currently available models have been validated against observed data.

Calculated low-dose risk estimates can give an impression of accuracy that is not justified, considering the assumptions and approximations on which they are based. For example, past risk assessments performed by the US Environmental Protection Agency have frequently focused on providing single-number upper-bound estimates of risk, which can be misleading. These assessments do not characterize either the uncertainty, *i.e.* the lack of knowledge as to the true situation, or the variability, *i.e.* the variation inherent in the parameters being measured. The routine use of mathematical models can mean that they are applied to incomplete or inappropriate bioassay data.

A further complication is that if QRA is used to calculate a single figure, which is then used to a set a legislative standard, the availability of new data or new QRA methods may lead to the setting of new and higher standards. An example is the setting of World Health Organization (WHO) guideline values for benzo[*a*]-pyrene in drinking water. Risk estimates made by WHO in 1984[5] and 1993[6] were based on the same data using different QRA methods, leading to recommendations for limits in drinking water which have risen from 0.01 μg l^{-1} (1984) to 0.7 μg l^{-1} (1993). The experimental data upon which both these estimates were based are of poor quality,[7] which would lead to uncertainty in the estimate even if the QRA methods were reliable. It is impossible to say whether this increase in the limit represents a significant increase in carcinogenic risk to consumers, but the 1984 guideline value almost certainly placed a financial burden on the providers of drinking water in some areas/countries and is now apparently regarded as unnecessary.

3　Setting Standards for Genotoxic Carcinogens Present in the Environment

While the overall UK approach is to strive to eliminate or reduce exposure to genotoxic compounds as far as possible, genotoxic carcinogens over which we have little control are present in the environment at low levels. These arise from human activities generally regarded as beneficial, including cooking and motor transport, and are usually at concentrations associated with a very low level of risk. The UK government is committed to using air quality standards for air pollutants, including carcinogens, and also sets maximum acceptable concentrations for carcinogens in drinking water. This does not mean that these standards are regarded as 'safe' concentrations, but scientific advice is that these are the concentrations at which the risk is regarded as extremely low and not measurable against background tumour incidences. Hence a pragmatic approach has been developed, based on the use of scientific expert judgement to define levels at which the risks to the general public would be expected to be exceedingly small and undetectable, even with sophisticated statistical techniques. It is then for government to consider both the benefit of the activity causing the contamination and the cost of reducing the contamination further, in deciding what action should be taken.

Premises upon which the UK Approach is Built

(a) "It is not possible to determine a meaningful No Observed Effect Level for genotoxic carcinogens, hence there is no level of exposure to a genotoxic carcinogen which should be described as 'acceptable' in health terms and in no need of further reduction".

(b) "Given that accurate estimation of the effects of (very low) ambient levels of genotoxic carcinogens upon the population is impossible, there is no case for allowing ambient levels of genotoxic carcinogens to rise". This is a very important premise since the calculation by QRA or other methods of a standard (if it is above the present ambient levels) could be regarded as 'a license to pollute'. If premise (a) is agreed, then this could never be acceptable, except in the most extreme case of public benefit from the activity causing the pollution or where to reduce the level of one carcinogen by a substantial amount the concentration of another would rise by a small amount.

(c) "Progressive reduction in ambient levels of genotoxic carcinogens is desirable. However, when levels are below the recommended standard (set using this approach), any further reductions should be undertaken with regard to the cost, both financial and social, and the feasibility of further reduction".

(d) "Based on the perception that prolonged, rather than short-term, exposure to chemical carcinogens is generally necessary to produce cancer in humans and that 'total exposure' is important, standards for the protection of human health should be defined in terms of a long averaging time, generally annual average exposure, since a lifetime average would be impossible to define". Since almost all available human exposure data are likely to relate to long-term occupational exposure, attempts to predict the effects of short-term exposures would introduce a further unquantifiable and undesirable uncertainty into a standard with a shorter averaging time.

Strategy for Developing an Air Quality Standard

A method of setting air quality standards, based upon these premises, scientific judgement, and the application of uncertainty factors has been described by Maynard *et al.*[8] This was broadly the approach taken by the UK Department of the Environment's Expert Panel on Air Quality Standards (EPAQS) when recommending a health-based air quality standard for benzene.[9]

The initial steps in the process are the assembly of a comprehensive database on the compound and deciding whether there are sufficient data of acceptable quality on which to base a decision on both the compound's carcinogenicity and the mechanism of carcinogenicity. In the case of benzene, this compound has been widely studied and is generally recognized as a genotoxic human carcinogen.

Defining a No Expected Human Effect Level (NEHEL)

Having decided that a compound is a potential human carcinogen with a genotoxic mechanism of carcinogenicity, the next step is to define a NEHEL if the data are sufficient. The defining of an NEHEL is a practical starting point for deriving an air quality standard for a genotoxic carcinogen: no assumption of a threshold of effect has been made.

It is accepted that a No Observed Effect Level cannot be defined for genotoxic carcinogens. However, it may be possible to define, from sound epidemiological studies, a level at which no effect would be likely to be demonstrable in any feasible epidemiological study. In defining this NEHEL both positive and negative findings are taken into account and expert judgement is used in assessing the quality and statistical power of the studies. An uncertainty factor may be included at the stage of defining the NEHEL to account for the uncertainties present in many retrospective epidemiological studies regarding exposure measurements, which are almost always incomplete, although it is generally assumed that they are correct within one order of magnitude.

It could be argued that if ever larger cohorts were studied, progressively lower NEHELs could be determined. However most studies of workers exposed to genotoxic carcinogens are retrospective studies, not least because occupational exposure to carcinogens is now strictly controlled to minimum levels. The opportunity to study large cohorts of workers exposed to varying, and some to high, concentrations of genotoxic carcinogens is declining and with it any possibility of defining the human exposure–response curve. For example a meta-analysis of studies of workers exposed to benzene,[10] covering a combined cohort of 208 000 petroleum workers, could not demonstrate a statistically significant increased risk of leukaemia in the combined cohort. Recent exposure levels in the petroleum industry have been much lower than those in the original studies that identified the link between benzene exposure and leukaemia.

Non-lymphocytic leukaemia was first described among shoe-factory workers exposed to very high ambient concentrations of benzene in Italy and Turkey. Subsequently, a series of epidemiological studies of workers in the synthetic rubber and petroleum industries, in which generally much lower exposures to benzene occurred, confirmed the increased risk of non-lymphocytic leukaemia due to exposure to benzene. None of these studies included satisfactory exposure estimates, because the exposures of interest occurred several decades ago, when monitoring methods were not well developed. Two studies provide usable estimates of exposures. The first was a cohort of 1165 workers at two rubber manufacturing plants between 1940 and 1965[11] (together with many subsequent refinings of the exposure estimates) and the second a cohort of 3536 chemical workers exposed to benzene between 1946 and 1975.[12,13] Both these studies showed an increased risk of non-lymphocytic leukaemias in workers with the highest exposures, estimated at greater than 200 ppm-years (*i.e.* equivalent to an average exposure of 10 ppm for 20 years) in the first study. In both studies there were excess cases of leukaemia in the lower exposure groups, and, although the increases in these groups were not statistically significant, there was a statistically significant trend between exposure to benzene and risk of developing leukaemia. Taking into account the uncertainties in the exposure estimates, a NEHEL was defined as 500 parts per billion (ppb) over a working lifetime. This is consistent with the judgement of the WHO task group on benzene,[14] who described 1 ppm as the exposure over a 40 year working career which had not been statistically associated with any increase in deaths from leukaemia. The use of judgement in the derivation of the NEHEL introduces an initial and unspecified uncertainty factor (comparison with the WHO conclusion suggests that in the case of benzene it is approximately 2), which increases the overall uncertainty factor in the eventual standard.

Developing a Standard from the NEHEL

Uncertainty factors are then applied to the NEHEL to account for differences between occupational and environmental exposure.

  (a) Divide by a factor of 10 to account for extrapolation of working life to whole life exposure (working lifetime is approximately 77 000 hours, while chronological lifetime is about 660 000 hours).
  (b) Divide by a factor of 10 to take into account distribution of sensitivity across the whole population, *e.g.* children and others who may have increased sensitivity to the compound.

This leads to a figure of X, expressed in appropriate units and as a long-term average.

Then the premise (b) on p. 47, that ambient levels of genotoxic carcinogens should not be allowed to rise, is introduced into the standard setting. If X is greater than the similarly defined ambient concentration A, then A is accepted as the standard; if X is lower than A then X is accepted as the standard.

It follows that for benzene X is defined as a 5 ppb running annual average. At the time when an air quality standard for benzene was recommended (1994) the annual average reported for central London was 4.2 ppb and levels rarely went above 5 ppb; current annual averages are 1–2 ppb. Therefore the standard was set at 5 ppb as a running annual average (see Figure 1).

When EPAQS[15] recommended an air quality standard for 1,3-butadiene (another genotoxic carcinogen present as an air pollutant from vehicle exhaust), a value of 10 ppb as a running annual average was defined as the equivalent of X. However concentrations of 1,3-butadiene in urban air do not exceed 1 ppb as a running annual average (equivalent to A) and so the recommended air quality standard for 1,3-butadiene was 1 ppb as a running annual average.

It has been suggested that this general approach could be extended to compounds for which acceptable data from epidemiological studies are not available.[8]

Use of a No Expected Animal Effect Level (NEAEL)

For some environmental contaminants judged to be genotoxic carcinogens there are insufficient or inadequate epidemiological data available to define a NEHEL. In this case the data from long-term animal bioassays have to be considered, and (if there are adequate data available) used in setting a standard. These data are used to define a NEAEL, *i.e.* the level at which one would not expect to demonstrate an effect in conventional long-term animal bioassays. If a range of studies are available in different species, then the results from studies in the most sensitive species should be used to define the NEAEL, unless there are strong mechanistic reasons (*e.g.* the effects depend upon metabolic pathways not present in humans) to discount these data. The NEAEL is then divided by two uncertainty factors of 10, one to account for interspecies variation and the other to account for the seriousness of the effect, *i.e.* cancer. The use of these factors of 10

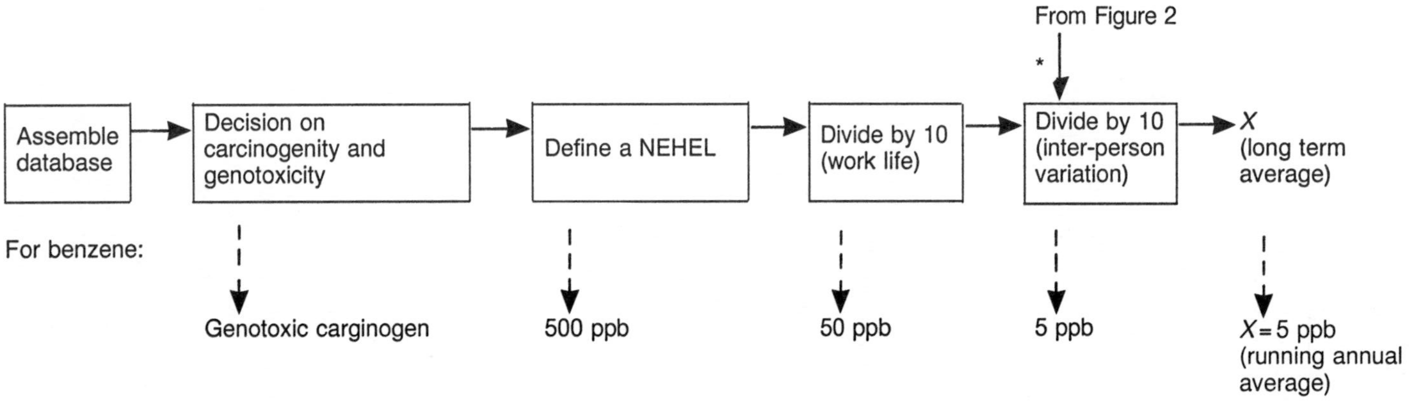

Is X greater than the equivalent ambient concentration, A?

If X is greater than A, then A = standard

If X is less than A, then X = standard

For benzene, both X and A = 5 ppb as a running annual average, therefore the standard = 5 ppb as a running annual average.

Figure 1 *Steps in setting an air quality standard, with benzene as an example*

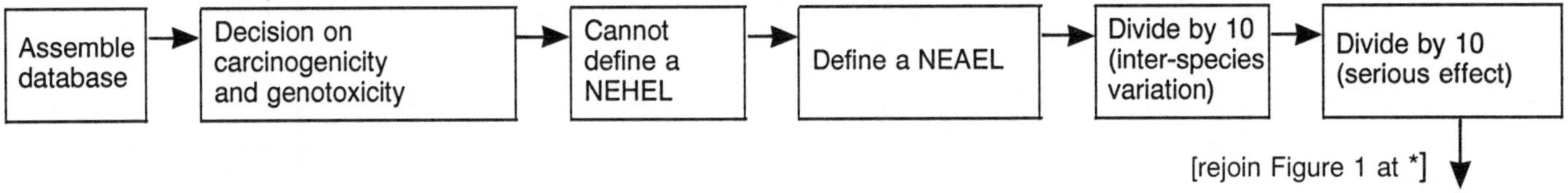

Figure 2 *Steps in the use of a No Expected Animal Effect Level (NEAEL)*

is arbitrary, although consistent with those used in the setting of acceptable daily intakes for compounds with threshold effects. The use of 10 as an uncertainty factor when extrapolating between species has been discussed by Renwick,[16] who suggested that when sufficient data become available regarding comparative toxicokinetics and toxicodynamics, then the uncertainty factor could be replaced with a scientifically derived correction factor that more accurately reflected the differences between the species.

This figure is then regarded as equivalent to the figure derived from epidemiological studies after application of one uncertainty factor of 10 (to account for lifetime exposure). It is therefore further divided by a factor of 10 to take into account distribution of sensitivity across the whole population, to give the figure for X (as a long-term average and in appropriate units), and the setting of the recommended standard continues as above (see Figure 2).

4   Standards and Targets

Although a standard for an environmental genotoxic carcinogen set in this way, if adhered to, will represent an exceedingly small risk to health,[9] the UK approach is to endeavour to keep exposure to such compounds as low as practicable. So a progressive reduction in environmental concentrations of genotoxic carcinogens is desirable. For air quality standards, for example, a lower target value can be recommended. However the date by which the target should be reached should only be set after consideration by policy-makers of cost and technological feasibility.

In the case of benzene a target value of 1 ppb as a running annual average has been recommended.[9] In the UK, projected levels of benzene emissions suggest that by the year 2000 exceedences of the 5 ppb standard will be virtually eliminated and very substantial progress made in reducing the extent of exceedences of the 1 ppb level.

5   Conclusions

An approach has been described for dealing with genotoxic carcinogens in the UK environment. This approach relies upon expert judgement at a number of points and in the UK this judgement is provided by a consensus opinion from expert advisory committees, generally made up of 10 to 20 medical and scientific experts. Because of this it is entirely possible that a different group of experts may, in particular, define the NEHEL or NEAEL differently. However the defining of these values by a group of experts who have considered all the available data guards against the likelihood of too much emphasis being placed upon a single study and allows a weighting of all the available studies.

Any method of setting standards that depends on scientific data should be open to revision in the light of new data, and regular re-examination will improve confidence in the standards. The reliance of the above strategy on the premise that exposure to genotoxic carcinogens should not be allowed to increase means that UK standard values for genotoxic carcinogens are unlikely to rise significantly (unless new information on genotoxic mechanisms comes to light). This also means that the introduction of new genotoxic carcinogens into the environment is discouraged, since if the present concentration is zero, then the recommended standard would also be zero.

This approach is a pragmatic means of setting standards for environmental genotoxic carcinogens, while being in accordance with other regulatory toxicological practice and sufficiently prudent to lead to a satisfactory level of protection of public health.

References

1 International Agency for Research on Cancer, 'IARC Monographs on the Evaluation of Carcinogenic Risks to Humans', IARC, Lyon, 1987, suppl. 7.
2 D. W. Gaylor, *J. Environ. Pathol. Toxicol.*, 1979, **3**, 179.
3 R. Peto, R. Gray, P. Brantom, and P. Grasso, *Cancer Res.*, 1991, **51**, 6415.
4 Department of Health, 'Guidelines for the Evaluation of Chemicals for Carcinogenicity', Report on Health and Social Subjects, HMSO, London, 1991, No. 42.
5 World Health Organization, 'Guidelines for Drinking Water Quality', WHO, Geneva, 1984.
6 World Health Organization, 'Guidelines for Drinking Water Quality', WHO, Geneva, 1993.
7 J. Neal and R. H. Rigdon, *Tex. Rep. Biol. Med.*, 1967, **25**, 553.
8 R. L. Maynard, K. M. Cameron, R. Fielder, A. McDonald, and A. Wadge, *Hum. Exp. Toxicol.*, 1995, **14**, 175.
9 Department of the Environment, Expert Panel on Air Quality Standards, 'Benzene', HMSO, London, 1994.
10 O. Wong and G. K. Raabe, *Regul. Toxicol. Pharmacol.*, 1995, **27**, 307.
11 R. A. Rinsky, A. B. Smith, R. Hornung, T. G. Fillon, R. J. Young, A. H. Okun, and P. J. Landrigan, *N. Engl. J. Med.*, 1987, **316**, 1044.
12 O. Wong, *Br. J. Ind. Med.*, 1987, **44**, 365.
13 O. Wong, *Br. J. Ind. Med.*, 1987, **44**, 382.
14 World Health Organization, 'Benzene', *Environmental Health Criteria*, WHO, Geneva, 1993, No. 150.
15 Department of the Environment, Expert Panel on Air Quality Standards, '1,3-Butadiene', HMSO, London, 1994.
16 A. G. Renwick, *Food Addit. Contam.*, 1993, **10**, 275.

Chapter 5

Dealing with Genotoxic Carcinogens: Refining the US Approach

By Bruce Molholt

ERM INC., 855 SPRINGDALE DRIVE, EXTON, PA 19341, USA

1 Introduction

Carcinogenesis is a multi-stage process requiring at least two genetic events in cellular oncogenes (*initiation*) followed by binary expansion of the mutant clone (*promotion*), and eventual further mutation to invade or metastasize (*progression*). Genotoxic carcinogens either mutate deoxyribonucleic acid (DNA) or alter chromosomes so as to cause heritable changes characteristic of either oncogenic initiation or progression. By contrast, the second stage of carcinogenesis, promotion, is engendered by agents not known to mutate DNA or alter chromosomes (non-genotoxic carcinogens). *Complete carcinogens* are those substances that are active in both initiation and promotion stages, *i.e.* possess both genotoxic and non-genotoxic activities (Figure 1).

Since each of these stages may take several years and burgeoning tumours are under constant attack by the immune system, frank carcinogenesis may not be realized until decades following exposure to initiating substances. Regardless of genotoxicity or carcinogenic mechanisms, the US Environmental Protection Agency (EPA) has mathematically modelled all carcinogens equivalently according to a Linearized Multi-Stage (LMS) equation for computing risk from exposure to known concentrations of carcinogens.

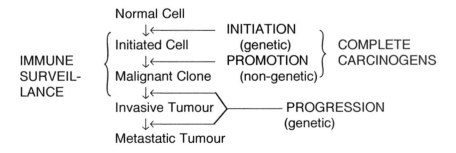

Figure 1

2 Complete Carcinogens

Very few chemical carcinogens are either completely initiators or completely promoters of carcinogenesis. As stated above, those substances with both genotoxic initiator and non-genotoxic promoter activities are referred to as complete carcinogens. Often chemicals that interact with DNA also interact with other cellular components, such as proteins at the cell surface, and are irritants at high doses. Irritation frequently is accompanied by hyperplasia, *i.e.* a proliferative response, which, if causing replication of initiated cancer cells, promotes carcinogenesis. Most often, this promotional activity in rodents is seen at the site of test substance administration, *i.e.* forestomach by ingestion and nasal epithelium by inhalation.

Three examples of complete carcinogens that are discussed in this paper primarily induce tumours at the site of administration, as noted in Table 1. Two of these, formaldehyde and propylene oxide, are gases at room temperature and, therefore, the normal route of exposure is inhalation. The other, benzo[*a*]pyrene, is an example of a polycyclic aromatic hydrocarbon and may be ingested as a product of combustion.

The carcinogenic mechanisms, including initiation and promotion activities, of each of the complete carcinogens shown in Table 1 is more fully described below.

Formaldehyde

Formaldehyde (HCHO) is a ubiquitous volatile compound to which all humans are exposed daily. Therefore it was of considerable concern when rats exposed to as little as 8 ppm HCHO in air came down with high proportions of nasal squamous cell carcinomas.[1,2] As an aqueous solution (formalin) HCHO also induces forestomach papillomas in male Wistar rats fed 0.5% formalin.[3] Hence, HCHO, like other complete carcinogens at maximally tolerated doses, shows tumours at the sites of administration (nasal epithelium by inhalation; forestomach by ingestion).

In addition, HCHO produces DNA–protein cross-links and is positively reactive in several short-term test systems for mutagenesis, including three bacterial genera, yeast, fungi, *Drosophila*, and mammalian cells in culture.[4]

Because HCHO is both genotoxic and irritating at high concentrations, its carcinogenicity in test animals exposed to high doses is a function of both

Table 1   *Examples of complete carcinogens*

Carcinogen	Exposure route	Site of carcinogenesis[a]
Formaldehyde	Inhalation	Nasal epithelium
Propylene oxide	Inhalation	Nasal vasculature
Benzo[*a*]pyrene	Ingestion	Forestomach

[a] Primary tumour site in experimental studies in rodents.

initiation and promotion activities.[5] This may be seen in the dose–response data generated by rats exposed by inhalation, as shown in Table 2. Obviously there is a disproportionate jump in tumour frequency between 5.6 ppm HCHO (1.3% tumours) to 14.3 ppm (67% tumours). Since the proportion of tumours is an order of magnitude greater at 14.3 ppm than predicted by the 5.6 ppm response, it may be concluded that the interaction of genotoxic and non-genotoxic activities at the higher dose causes a synergistic effect on the number of squamous cell carcinomas induced.

Propylene Oxide

Like the more commonly encountered ethylene oxide, propylene oxide is a highly reactive epoxide and a strongly irritating gas at high concentrations in air. Unlike benzo[a]pyrene (below), neither ethylene oxide nor propylene oxide require metabolic activation to form mutagenic DNA adducts. An extrapolation based upon relative surface areas of laboratory rodents and humans is not required for these direct-acting epoxides. Hence, carcinogenic dose–response data from rodents may be utilized (in mg per kg of body weight) for the estimation of potential risks for human receptors who have been exposed to propylene oxide.[6]

The US National Toxicology Program (NTP) studies in 1985[7] have been utilized to establish the inhalation carcinogenic potency of propylene oxide in chronically exposed mice based upon the induction of nasal hemangiomas and hemangiosarcomas. The data obtained are shown in Table 3.

Clearly, a threshold for propylene oxide inhalation carcinogenesis exists in mice between 200 and 400 ppm (55 and 110 mg kg^{-1} day^{-1}). If carcinogenesis at 200 ppm was proportional to the 400 ppm response, then approximately 5/50

Table 2   *Nasal tumours in rats from HCHO inhalation*[a]

HCHO dose	Tumour incidence	Percent
0 ppm (control)	0/156	0.00
2.0 ppm	0/159	0.00
5.6 ppm	2/153	1.31
14.3 ppm	94/140	67.1

[a] Data from reference 5.

Table 3   *Inhalation carcinogenesis by propylene oxide in mice*[a]

Dose (ppm)	Sex	Hemangio(sarco)mas
0	M	0/50
	F	0/50
200	M	0/50
	F	0/50
400	M	10/50
	F	5/50

[a] Data from reference 7.

male mice and 2–3/50 female mice would have developed nasal tumours at this dose. Hence, the carcinogenic response at 400 ppm is most likely the product of initiation and promotion activity of propylene oxide at irritating doses. As in the case of ingested benzo[*a*]pyrene promotion in forestomach, these hemangiomas and hemangiosarcomas in mice inhaling high concentrations of propylene oxide were all seen in nasal epithelium, the site of administration.

Benzo[*a*]pyrene

Benzo[*a*]pyrene is a coplanar collection of five benzene rings and prototypic of the carcinogenic polycyclic aromatic hydrocarbons (PAHs). In an effort to solubilize PAHs, liver enzymes add hydroxyl ($-OH$, alcohol, 'ol') or epoxide (–O–) groups to select carbon atoms, which provide sites for benzene ring breakage and solubilization, such that these toxic molecules may be excreted in urine. Unfortunately, a diol epoxide metabolite of benzo[*a*]pyrene (BPDE) is active in carcinogenic initiation. The highly reactive epoxide moiety of BPDE covalently bonds with the N7 group of guanine, which adduct when replicated may lead to a guanine→thymine transversion point mutation. However, it is clear from many experiments with unmetabolized benzo[*a*]pyrene that it is irritating at high doses and induces many enzymes and cellular responses that are consistent with carcinogenic promotion.

The 1967 data of Neal and Rigdon[8] have been utilized by the US EPA to derive an ingestion carcinogenic potency for benzo[*a*]pyrene. Since irritation normally occurs at the first point of tissue contact, Neal and Rigdon monitored tumours of the forestomach in mice fed varying concentrations of benzo[*a*]pyrene of up to 250 ppm daily for 2 years. Their results clearly show a threshold for forestomach carcinogenesis, as seen in Table 4.

It may be seen that benzo[*a*]pyrene-induced forestomach carcinogenesis in mice begins abruptly between 40 and 50 ppm. This threshold is very similar to that seen for another carcinogenic PAH, 2-acetylaminofluorene, in the induction of bladder cancers in mice, as will be discussed further below.

Table 4 *Mouse forestomach tumours after feeding benzo*[a]*pyrene*[a]

Dose (ppm)	Tumour incidence	Tumour frequency (%)
0	0/289	0
1	0/25	0
10	0/24	0
20	1/23	4.3
30	0/37	0
40	1/40	2.5
45	4/40	10.0
50	24/34	70.6
100	19/23	82.6
250	66/73	90.4

[a] Data of reference 8.

3 *In Vitro* Assays for Mutagenesis

In general, cells in culture (*in vitro*) may behave quite differently from cells within the body of a living organism (*in vivo*). Whereas *in vitro* mutagenic assays may be good at distinguishing genotoxic from non-genotoxic carcinogens, in that replicative controls, DNA repair, and other rate-limiting phenomena are less constrained outside the body, they are less predictive of mutagenic potencies.[9] Hence, *in vivo* assays for mutagenesis are emphasized below.

4 *In Vivo* Assays for Mutagenesis

Transgenic Mice

The insertion of foreign genes into embryologic mice results in transgenic mice in which every cell, including the germline, carries one or more copies of the foreign gene. Extrapolation of mutagenic data from transgenic mice to humans seems straightforward based upon an analysis of spontaneous mutations in the human factor IX gene.[10]

In Japan, Gondo *et al.*[11] have created 20 transgenic mouse strains that utilize a shuttle plasmid vector encoding bacterial streptomycin resistance. In the US and other western countries, bacterial genes in the lactose operon (*lacZ*, encoding β-galactosidase, and *lacI*, encoding the lactose repressor) have proved useful in quantifying mutagenic potential. These are not required genes for survival in transgenic mice and, therefore, their mutagenesis does not inhibit cell survival. So far 26 chemicals have been evaluated for mutagenesis in *lacZ*- or *lacI*-bearing tissues in eight organ sites of transgenic mice.[12]

HPRT and Other Genetic Tests

The hypoxanthine–guanine phosphoribosyl transferase (HPRT) gene (*hprt*) is encoded upon the X-chromosome and is, thus, haploid in males. Cells mutant for HPRT are resistant to 6-thioguanine and easily scored. The background frequency of HPRT mutations is 5×10^{-6}. This locus has been widely employed as a short-term test system in V79 Chinese hamster cells[13] and human circulating T-lymphocytes.[14] It is especially interesting that *hprt* mutants in human T-lymphocytes frequently involve illegitimate chromosomal translocations mediated by the V(D)J recombinase system normally utilized in generating immunoglobulin diversity.[15] This mechanism may be especially important in leukemogenesis and lymphomagenesis and is seen in 40% of the spontaneous *hprt* mutations in neonatal T-lymphocytes.[15]

The *hprt* gene has also been used to characterize the molecular nature of mutagenic lesions induced by carcinogenic initiators in human cells. For example, aflatoxin B1 induces up to 1/6 *hprt* mutants at a mutational hotspot in base 209 of exon 3.[16] This hotspot is in a hexanucleotide guanine (G)(GGGGGG) stretch and results in a Gcytosine(C)→thymine(T)adenine(A) tranversion.

DNA Adducts

Many genotoxic carcinogens cause the formation of DNA adducts as intermediates in mutagenesis. Benzo[*a*]pyrene and other carcinogenic PAHs are metabolized to DNA base-reactive epoxides by cytochrome P-450 enzymes.[17] Propylene oxide and other epoxides act directly on DNA to form adducts without the need for metabolic activation.[6] These DNA adducts may be used as biomarkers for human exposure to genotoxic carcinogens.[18]

Single-strand DNA Breaks

Carcinogenic initiation damage to DNA is reparable in a multi-step process involving excision and replacement of damaged bases in one DNA strand using the complementary strand as a template. Hence, the potency of genotoxic carcinogens may be quantified by the number of single-strand breaks introduced by DNA repair following carcinogen exposure. Steeb *et al.*[19] have utilized this assay to differentiate genotoxic and non-genotoxic carcinogens and to quantify genotoxic potencies.

5 Dose–Response Relationships for Genotoxic Carcinogens

The EPA's LMS model appears to accurately predict the carcinogenicity of some direct-acting genotoxic carcinogens for several orders of magnitude in the dose–response curve. For example, the mouse skin assay for papilloma induction by benzo[*a*]pyrene plus the phorbol ester TPA shows a linear response from about 100 tumours per mouse down to 1 tumour per hundred mice (Figure 2). Such linearity also appears with another PAH initiator, 2-acetylaminofluorene (2-AAF) in mouse liver, albeit over less than two orders of magnitude in response over background (Figure 3). In mouse bladder, however, the dose–response curve is distinctly non-linear with a threshold for bladder carcinogenesis at 50 ppm (Figure 3).

Close examination of the liver dose–response curve for 2-AAF shows that it, too, has a positive inflection above 50 ppm. A logical interpretation of these data is that 2-AAF initiates carcinogenesis in the liver, which is rich in metabolic enzymes, but not in the bladder. The dose–response curve below 50 ppm in liver reflects purely carcinogenic initiation, above 50 ppm in bladder purely carcinogenic promotion, and above 50 ppm in liver both initiation and promotion.

6 Dose–Response Relationships for Non-genotoxic Carcinogens

The vast majority of carcinogens regulated by the EPA are not genotoxic. Most of these are positive for carcinogenicity only at high doses in the B6C3F1 strain of

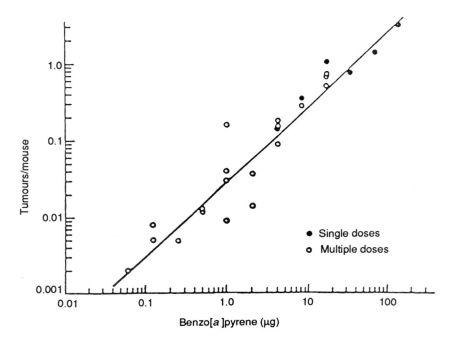

Figure 2   *Dose–response curve for benzo[a]pyrene induction of tumours in mouse skin*

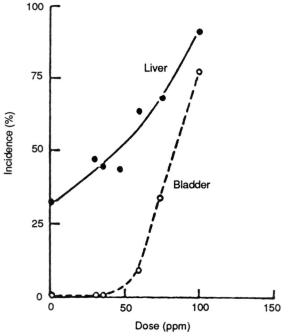

Figure 3   *Dose–response curve for 2-acetylaminofluorene induction of tumours in mouse liver and bladder*

mouse and induce solely liver tumours. Non-genotoxic chemicals do not initiate carcinogenesis, but only promote or immunosuppress. Like the dose–response curve for 2-AAF in bladder carcinogenesis (Figure 3), the dose–response curves for pure promoters have threshold concentrations below which carcinogenesis is not exacerbated (see reviews in references 22–25).[20–25]

2,3,7,8-Tetrachlorodibenzodioxin is a much studied non-genotoxic carcinogen and displays a non-linear dose–response curve for carcinogenesis.[26–29]

7 Dose–Response Relationships for Complete Carcinogens

The dose–response curve for benzo[*a*]pyrene in the induction of forestomach tumours in rats is a useful prototype for demonstration of the sigmoid shape of the dose–response curve for complete carcinogens (Figure 4). During the steepest portion of the curve between 40 and 50 ppm, both initiation and promotion activities are operational, whereas below 40 ppm promoter activity atrophies and a new, much lower slope reflects initiation activity alone. The slope of the initiation curve is 0.04% gastric tumours per ppm, whereas the slope of the initiation × promotion curve is 8.0. If the assumption is made that environmental exposures to benzo[*a*]pyrene are normally within the range below 40 ppm, then use of the higher slope value exaggerates actual risk by 200-fold.

8 Mutagenic Potencies for Genotoxic and Complete Carcinogens

As a test for the thesis described herein, it is useful to compare mutagenic and carcinogenic potencies (MP and CP, respectively) where data are available. Pure initiators, *i.e.* those carcinogens with solely genotoxic activity, should have MP/CP ratios that are much higher than complete carcinogens, while pure promoters should have MP/CP ratios ≈ 0. Meselson and Russell[30] have attempted to rank several carcinogens, including three aromatic amines, according to their comparative mutagenic and carcinogenic potencies. Table 5 is derived from their data.

Since the analysis (Table 5) was conducted with mutagenic data in microbial cells in the presence of cytochrome P-450 activation, the conversion of procarcinogens to ultimate carcinogens, *i.e.* metabolism to initiator species, was maximal. This may be an exaggeration of initiator/promoter activities for these compounds *in vivo*. The three aromatic amines have MP/CP ratios, *i.e.* evidence for carcinogenic initiation activities, which are an order of magnitude above several well-known strong carcinogenic initiators. The most straightforward interpretation of these data is that the aromatic amines active in bladder carcinogenesis (benzidine, 4-aminobiphenyl and β-naphthylamine) are those carcinogens that act primarily by initiation rather than promotion. MMS and NMU, both classic mutagens, appear to irritate as much as initiate at doses that

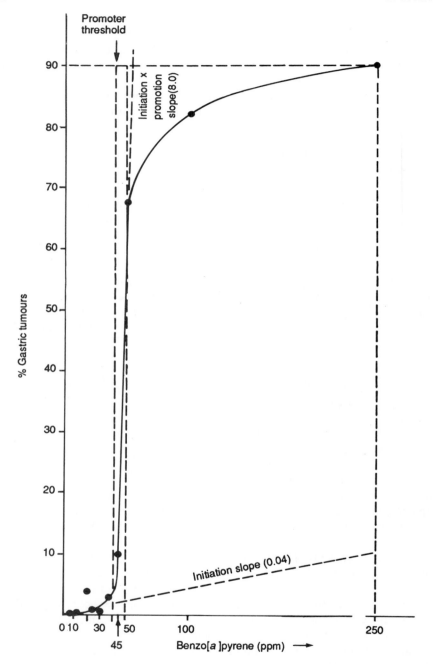

Figure 4   *Dose–response curve for benzo[a]pyrene induction of tumours in mouse forestomach*

Table 5 *Mutagenic and carcinogenic potencies for selected compounds*[a] (Modified from reference 30)

Compound	Mutagenic Potency (MP)	Carcinogenic Potency (CP)	MP/CP
Aflatoxin B1	23	820	0.28
Benzo[a]pyrene	4.8	5.5	0.87
Dibenz[a,h]anthracene	0.40	0.47	0.85
Methylmethane sulfonate (MMS)	0.057	0.032	1.8
N-Nitrosomethylurea (NMU)	0.43	0.10	4.3
Benzidine	0.53	0.056	9.5
4-Aminobiphenyl	1.8	0.18	10
β-Naphthylamine	0.59	0.028	21

[a] Mutagenic and carcinogenic potencies calculated from *Salmonella his*[+] revertants in the Ames test complete with cytochrome P-450 activation and rodent lifetime bioassays by injection or ingestion, respectively.

have been used experimentally in rodent carcinogenic assays. The three PAH carcinogens tabulated, aflatoxin B1, benzo[a]pyrene, and dibenz[a,h]anthracene, are not mutagenic *per se*, but, also like the aromatic amines, require metabolic activation to reactive species that adduct DNA at guanine residues.

9 Synergistic Interactions in Complete Carcinogenesis

The risks from combinations of genotoxic and non-genotoxic carcinogens is known to be synergistic, *i.e.* greater than the sum of the risks from individual substances. For example, in the seminal epidemiologic studies of Selikoff *et al.*[31], whereas lung cancer risks were found to be 10-fold greater from smoking and 10-fold greater from asbestos exposure, the combination of smoking and asbestos exposure was found to be 100-fold, rather than the 20-fold additive risk anticipated. This same synergy is apparent for complete carcinogens at doses above promoter thresholds where genotoxic and non-genotoxic effects are multiplicative.

A combination of short-term tests that are sensitive to both genotoxic and non-genotoxic activities of complete carcinogens appears more predictive of carcinogenic risk than the normally utilized battery of test systems.[9]

10 Risk Assessment Methodology for Complete Carcinogens

Mutagenesis may be assessed most easily in short-term test systems. Whether microbial or mammalian cell cultures, these assays can detect mutagenesis inexpensively and within a matter of days. However, these *in vitro* systems are not particularly good at replicating the *in vivo* environment in terms of activating enzymes, DNA repair rates, *etc.* Hence, it is suggested that transgenic mice, *hprt* mutations in circulating lymphocytes, DNA adducts, or other *in vivo* end-points be employed to measure mutagenic potential not only qualitatively but also quantitatively such that mutagenic potency factors (MPFs) might be derived for each complete carcinogen.

As with the conventional EPA risk assessment paradigm, it is suggested that MPFs be expressed in inverse mg of substance per kg of body weight per day [(mg kg^{-1} $day^{-1})^{-1}$], such that lifetime carcinogenic risk from exposure to low levels of complete carcinogens be expressed as:

$$Risk = CDI \times MPF$$

where Risk = the lifetime risk for carcinogenic initiation; CDI = chronic daily intake in mg kg^{-1} day^{-1}; MPF = mutagenic potency factor in (mg $kg^{-1} day^{-1})^{-1}$.

11   Conclusions

In the US, the potential risk from exposure to complete carcinogens is assessed via a LMS model which predicts human risks at low exposures from experimental data derived from animals that have been exposed to maximally tolerated doses. In that many genotoxic carcinogens are irritating and promote carcinogenesis at high doses, this US methodology frequently overestimates risks to humans at doses that are most likely to be encountered in the environment.

Whereas the LMS model is accurate in predicting the frequency of carcinogenesis at low doses of initiating carcinogens, the available evidence points towards the existence of thresholds below which complete carcinogens fail to promote. Hence, a model is required that accurately predicts complete carcinogen risks beneath promoter threshold doses. Rather than utilizing the EPA's LMS model, it is proposed that potential risks from ambient exposures to complete carcinogens be modelled according to mutagenic potency at low doses rather than carcinogenic potency at high doses.

Glossary of Terms

Ambient Levels

Concentrations of chemicals in soil, water, or air that are likely to be encountered in the environment as the result of industrial activity.

BPDE

Benzo[*a*]pyrene diol epoxide: the metabolite of benzo[*a*]pyrene that covalently bonds with guanine, forming a mutagenic DNA adduct.

Carcinogenesis

A clonal multi-step process that converts normal cells into tumours.

Chromosomes

Collections of DNA and proteins that encode genetic information. In the human cell, there are 22 pairs of autosomal chromosomes plus XX (female) or XY (male) making 46 chromosomes in all.

Clastogenic

Causing chromosomal breaks, deletions, or rearrangements.

Complete Carcinogen

Having both genotoxic (initiating) and non-genotoxic (promoting) carcinogenic activities.

DNA

Deoxyribonucleic acid: the double-stranded helical ensemble of alternating deoxyribose and phosphate molecules on the outside connected by paired bases (adenine: thymine; guanine: cytosine) on the inside which encodes the sequence of amino acids for protein synthesis.

Genotoxic

Mutagenic, inducing mutations in DNA, or clastogenic, inducing chromosomal breaks, deletions, or translocations.

HPRT

Hypoxanthine–guanine phosphoribosyl transferase: a key enzyme in the pathway for purinoside biosynthesis. Mutants in the gene *hprt* allow cells to thrive in the otherwise lethal purine analogue 6-thioguanine.

Immunosuppression

Suppression of the immune system, which, in terms of carcinogenesis, means specific suppression of those T-lymphocytes, including cytotoxic NK-cells, that are naturally active in rejection of tumour cells.

In Vitro

Outside the body (literally 'in glass'); referring here to experiments conducted with metazoan cells such as lymphocytes perpetuated in cell culture.

In Vivo

Inside the body (literally 'in life'); referring here to experiments conducted within metazoan creatures, including experimental animals and humans.

Initiation

Used here to describe the molecular process that first transforms normal cells to genetically altered precursors which predispose them to carcinogenesis.

LMS Model

The linearized multi-stage model of dose–response developed by the US Environmental Protection Agency to assess the potential lifetime risk from exposure to carcinogens.

MTD

Maximally tolerated dose; the quantity of a given chemical just below that which is lethal to 50 percent of exposed test animals.

Mutagenic

Causing mutations, *i.e.* base substitutions, deletions, or rearrangements in DNA.

Non-genotoxic

Not causing mutations or chromosomal damage/rearrangement when assessed by short-term tests. Non-genotoxic carcinogens are active in promotion or immunosuppression.

PAH

Polycyclic aromatic hydrocarbon (or polyaromatic hydrocarbon): that class of fused benzene ring compounds including benzo[a]pyrene, 2-acetylaminofluorene, and other carcinogenic compounds found in combusted materials.

Promotion

The second stage of carcinogenesis, following initiation, in which initiated stem cells are stimulated to proliferate. Promoters are not genotoxic.

Risk Assessment

Assessment of potential risk from exposure to hazardous substances; utilized here to describe paradigms for assessing potential carcinogenic risks from exposure to genotoxic, non-genotoxic, or complete carcinogens.

Short-term Tests

Used here as tests for genotoxicity that are performed in microbial or mammalian cell cultures with end-points that may be measured in days.

Synergy

Cooperativity, as used here, is explicitly that cooperative interaction between carcinogenic initiator and promoter activities that results in multiplicative rather than additive exacerbation of risk.

Transgenic Mice

Mice that have been genetically altered to contain foreign genes in every cell.

References

1 T.B. Starr and J.E. Gibson, 'The mechanistic toxicology of formaldehyde and its implications for quantitative risk estimation', *Annu. Rev. Pharmacol. Toxicol.*, 1985, **25**, 745–767.

2 H.M. Bolt, 'Experimental toxicology of formaldehyde', *J. Cancer Res. Clin. Oncol.*, 1987, **113**, 305–309.

3 M. Takahashi, R. Hasegawa, F. Furukawa, K. Toyoda, H. Sato, and Y. Hayashi, 'Effects of ethanol, potassium metabisulfite, formaldehyde and hydrogen peroxide on gastric carcinogenesis in rats after initiation with *N*-methyl-*N*'-nitro-*N*'-nitrosoguanidine', *Jpn. J. Cancer Res.*, 1986, **77**, 118–124.

4 A.G. Ulsamer, K.C. Gupta, M.S. Cohn, and P.W. Preuss, 'Formaldehyde in indoor air: Toxicity and risk', in Proc. 75th Annu. Meet. Air Pollut. Control Assoc., 1982, 16 pp.

5 W.D. Kerns, K.L. Pavkov, D.J. Donofrio, E.J. Gralla, and J.A. Swenberg, 'Carcinogenicity of formaldehyde in rats and mice after long-term inhalation exposure', *Cancer Res.*, 1983, **43**, 4382–4392.

6 D. Segerbäck, S. Osterman-Golkar, B. Molholt, and R. Nilsson, '*In vivo* tissue dosimetry as a basis for cross-species extrapolation in cancer risk assessment of propylene oxide', *Regul. Toxicol. Pharmacol.*, 1994, **20**, 1–14.

7 US National Toxicology Program, 'Toxicology and carcinogenesis studies of propylene oxide (CAS No. 75-56-9) in F344/N rats and B6C3F1 mice (inhalation studies)', US NTP, National Institutes of Health, Publ. No. 85-2527, 1985.

8 J. Neal and R.H. Rigdon, 'Gastric tumors in mice fed benzo[*a*]pyrene: A quantitative study', *Tex. Rep. Biol. Med.*, 1967, **25**, 553–557.

9 K.T. Kitchin, J.L. Brown, and A.P. Kulkarni, 'Complementarity of genotoxic and nongenotoxic predictors of rodent carcinogenicity', *Teratog. Carcinog. Mutagen.*, 1994, **14**, 83–100.

10 S.S. Sommer and R.P. Ketterling, 'How precisely can data from transgenic mouse

mutation-detection systems be extrapolated to humans? Lessons from the human factor IX gene', *Mutat. Res.*, 1994, **307**, 517–531.

11 Y. Gondo, Y. Ikeda, Y. Motegi, A. Takeshita, M. Takahashi, K. Nakao, and M. Katsuki, 'Development of new risk assessment test by using transgenic mice', *Kankyo Hen'igen Kenkyu*, 1994, **16**, 53–65.

12 V. Morrison and J. Ashby, 'A preliminary evaluation of the performance of the Muta Mouse (*lacZ*) and Big Blue (*lacI*) transgenic mouse mutation assays', *Mutagenesis*, 1994, **9**, 367–375.

13 H. Glatt, 'HPRT-gene mutation test in V79-cells of Chinese hamsters', *Mutat. Genet. Toxicol.*, 1993, 243–262.

14 W. G. McGregor, V. M. Maher, and J. J. McCormick, 'Kinds and locations of mutations induced in the hypoxanthine–guanine phosphoribosyl transferease gene of human T-lymphocytes by 1-nitrosopyrene, including those caused by V(D)J recombinase', *Cancer Res.*, 1994, **54**, 4207–4213.

15 Hou, S-M., 'Novel types of mutation identified at the hprt locus of human T-lymphocytes', *Mutat. Res.*, 1994, **308**, 23–31.

16 N. F. Cariello, L. Cui, and T. R. Skopek, '*In vitro* mutational spectrum of aflatoxin B1 in the human hypoxanthine guanine phosphoribosyl transferase gene', *Cancer Res.*, 1994, **54**, 4436–4441.

17 A. H. Conney, 'Induction of microsomal enzymes by foreign chemicals and carcinogenesis by polycyclic aromatic hydrocarbons', *Cancer Res.*, 1982, **42**, 4875–4917.

18 P. B. Farmer, 'Carcinogen adducts: Use in diagnosis and risk assessment', *Clin. Chem.*, 1994, **40**, 1438–1443.

19 L. Steeb, H. Friesel, T. Schneider, and E. Hecker, 'Potential tumor initiating agents for skin: Binding to DNA and induction of DNA single-strand breaks measured by nick translation', *Polycyclic Aromat. Cmpd.*, 1993, **3** (Suppl), 857–862.

20 F. J. Burns and R. E. Albert, 'Relationship of benign papillomas to cancer induction in mouse skin', *Cancer Detect. Prev.*, 1981, **4**, 99–107.

21 N. A. Littlefield, J. H. Farmer, D. W. Gaylor, and W. G. Sheldon, 'Effects of dose and time in a long-term, low-dose carcinogenic study', *J. Environ. Pathol. Toxicol.*, 1979, **3**, 17–34.

22 B. E. Butterworth, 'Consideration of both genotoxic and nongenotoxic mechanisms in predicting carcinogenic potential', *Mutat. Res.*, 1990, **239**, 117–132.

23 E. Hecker and F. Rippmann, 'Outline of a descriptive general theory of environmental chemical cancerogenesis. Experimental threshold doses for tumor promoters', in 'Mechanisms of Environmental Mutagenesis and Carcinogenesis', ed. A. Kappas, Plenum Press, New York, 1990, pp.173.

24 H. C. Pitot and Y. P. Dragan, 'Facts and theories concerning the mechanisms of carcinogenesis', *FASEB J.*, 1991, **5**, 2280–2286.

25 J. H. Weisburger and G. M. Williams, 'Types and amounts of carcinogens as potential human cancer hazards', *Adv. Mod. Environ. Toxicol.*, 1990, **17**, 209–223.

26 D. G. Barnes, 'Toxicity equivalents and EPA's risk assessment of 2,3,7,8-TCDD', *Sci. Total Environ.*, 1991, **104**, 73–86.

27 R. E. Keenan, D. J. Paustenbach, R. J. Wenning, and A. H. Parsons, 'Pathology reevaluation of the Kociba *et al.* bioassay of 2,3,7,8-TCDD: Implications for risk assessment', *J. Toxicol. Environ. Health*, 1978, **34**, 279–296.

28 eds. M. A. Gallo, R. J. Scheuplein, and K. A. van der Heijden, 'Biological Basis for Risk Assessment of Dioxins and Related Compounds', Banbury Report 35, Cold Spring Harbor Press, NY, 1991.

29 US Environmental Protection Agency, 'Health assessment document for 2,3,7,8-

tetrachlorodibenzo-*p*-dioxin (TCDD) and related compounds', Office of Research and Development, US EPA, Washington, D.C., 1994, External review draft EPA/600/BP-92/001.

30 M. Meselson and K. Russell, 'Comparisons of carcinogenic and mutagenic potency', in 'Origins of Human Cancer', eds. H. H. Hiatt, J. D. Watson, and J. A. Winsten, Cold Spring Harbor Laboratory, New York, 1977, pp. 1473–1481.

31 I. J. Selikoff, E. C. Hammond, and J. Churg, 'Asbestos exposure, smoking and neoplasia', *J. Am. Med. Assoc.*, 1968, **204**, 106–110.

Chapter 6

Epidemiological Investigation of Environmental Health Issues

By Stuart Pocock

DEPARTMENT OF EPIDEMIOLOGY AND POPULATION SCIENCES,
LONDON SCHOOL OF HYGIENE AND TROPICAL MEDICINE,
LONDON WC1E 7HT, UK

1 Introduction: The Value of Environmental Epidemiology

Much valuable information on chemical and other environmental hazards can be obtained by toxicological studies in the laboratory. Animal experiments and other techniques such as mutagenic testing are very useful in identifying specific chemical hazards, given the tightness of scientific rigour that can be achieved in laboratory experiments. However, such studies may only be of indirect relevance to human exposure. Animal models, especially at high doses, cannot give clear guidelines to the human situation, which is often concerned with much lesser exposures affecting substantial populations. Thus, toxicology provides a valuable rationale for *formulating hypotheses* about potential environmental hazards. The epidemiological approach then provides the means to assess the reality of such hazards in the human population.

The directness of epidemiology in studying actual human exposure would seem its great strength. However, scientific influences are most readily drawn from data in controlled experimental studies, whereas environmental epidemiology is inevitably based on an observational approach. This limitation of the epidemiological method will be discussed more fully in Section 9, together with the difficulty in making policy decisions from epidemiological evidence. Firstly we must define the main issues to be considered when undertaking studies in environmental epidemiology. These issues are covered in Sections 2 to 8, as follows: the type of study design, measures of disease outcome, measures of environmental exposure, confounding factors, measurement error, avoidance of bias, size of study, and the principles of statistical analysis.

2 Types of Epidemiological Study

There exist several observational approaches to studying potential environmental hazards in the human population.

Prospective Studies

Here one identifies a cohort of subjects for whom some individual marker(s) of exposure can be obtained. One then follows up each subject to see which relevant measures of ill health subsequently occur. Such prospective (longitudinal, cohort) studies have been especially valuable in the occupational environment (*e.g.* in studies of cancer and causes of death in asbestos workers) where definition of the relevant study sample, measures of individual exposure and long-term follow-up are more feasible than in studies in the general community. Even so, the burden of follow-up makes such studies costly and lengthy to perform. Such a longitudinal approach has recently been adopted in studies of body lead burden and neuropsychological development in birth cohorts.

Case-Control Studies

Here one identifies a sample of subjects (cases) with a given disease outcome and compares them with a sample of subjects (controls) without the disease, to see if the cases have a higher exposure to the environmental hazard of interest. Such case-control studies are of limited use in environmental studies since retrospective identification of individual exposure is often impossible. They would be most appropriate in situations where an uncommon disease may plausibly be related to a particular environmental hazard (*e.g.* mesothelioma and asbestos). They have also been used in relating environmental lead to (i) mental retardation and (ii) hypertension, but in fact cross-sectional studies (see below) of (i) mental performance, and (ii) blood pressure, are perhaps more useful.

Cross-Sectional Studies

Here one identifies an appropriate sample of subjects and relates current indicators of ill health to a current (or past) measure of exposure. Often one aims to include subjects with a sufficient spread of exposure so that a valid international assessment of the possible exposure/ill health association can be achieved. Such a cross-sectional approach is useful in studying the possible effects of environmental and occupational hazards on chronic adult conditions (*e.g.* respiratory disease, hypertension) and has also been used extensively in studies of lead and child neuropsychological performance at around age 6. The main drawbacks concern the difficulty of getting valid measures of earlier exposure and, in occupational studies, the potential withdrawal from the workforce of subjects most severely affected.

Geographic Studies

For many potential environmental hazards (*e.g.* contaminants in water supplies) it is difficult to obtain measures of exposure in individual subjects. Here one undertakes geographic comparisons by obtaining for a sample of well-defined communities both a measure of community exposure and community measures of disease. Since one is usually dependent on available official statistics, such geographic studies are often based on disease-specific mortality rates or cancer incidence rates. Geographic studies can also be concerned with *time trends* in disease, *e.g.* the recent study relating changes in perinatal mortality in different German regions to fallout from Chernobyl. Such geographic studies are fraught with problems, in both their conduct and their interpretation, especially as regards obtaining reliable measures of community exposure (particularly *past* exposure) and the fact that substantial community differences in disease will exist for reasons other than the exposure of prime interest.

Point Source Clustering Studies

In studying the effects of a specific local environmental source (*e.g.* a chemical plant or a nuclear installation) one is often concerned with relating disease events (especially childhood cancers) to distance from the point source. In recent times there have been considerable methodological advances for such studies. For instance, instead of being confined to disease rates within local administrative areas a more defined investigation of potential clustering can be obtained by relating individual events and corresponding populations at risk to distance from the point source. Also, the collection of more refined small area health statistics (mortality and cancer incidence) on a national basis can lead to more general studies of disease clustering with the aim of actively seeking previously unidentified environmental hazards.

3   Measures of Disease Outcome

Studies in environmental epidemiology are more likely to reach clear conclusions when there is a well-defined aetiological hypothesis, relating exposure to a single specific and measurable disease outcome, which has a sensible biological basis derived from toxicological studies. Problems occur if:

(a) the disease outcome is unreliably measured, *e.g.* a recent study of aluminium in water and Alzheimer's disease was hindered by the potential inconsistency in determining Alzheimer's cases from brain scan reports.
(b) there is no established measure of disease outcome, *e.g.* in studies relating lead to child neuropsychological development there is considerable uncertainty over what neuropsychological aspects may plausibly be affected by low level exposure and how to measure those aspects.
(c) several disease outcomes are studies in a 'fishing expedition' which has no clear *a priori* aetiological priorities. For instance, in a recent study of point source clustering around a chemical plant numerous cancer sites were analysed in both adults and children. The consequent focus on an excess of cancers of the central nervous system was data-provoked, so that while such exploratory data analysis is of interest in formulating hypotheses for

future study, great caution is required in interpreting any *post hoc* emphasis on positive findings.

The appropriate measure(s) of disease outcome need to be clearly defined in advance. Types of outcome may be broadly classified into:

(a) *mortality*, in which the usual approach is to determine each death's underlying cause according to the standard rules of the International Classification of Disease. The reliability of centralized national tracing and recordings of deaths varies between countries, and the inevitable errors of diagnosis, particularly in the elderly, need to be borne in mind.

(b) *incidence*, in which new cases of a disease are identified at diagnosis. This is most commonly used in cancer studies, where formalized records of site-specific cancers are kept in many countries. Cancer incidence can be more useful than cancer mortality, because it is nearer in time to the environmental exposure of interest and also avoids the problem of varying case survival. Incidence data for other diseases are usually hard to obtain with any reliability, except by extensive follow-up of individual subjects, which is usually unrealistic except possibly in the occupational setting.

(c) *prevalence*, in which subjects are all examined on one or more occasions to determine whether disease is present. Prevalence data may consist of:
   (i) the presence/absence of a disease condition, *or*
   (ii) an ordered classification where each individual is graded according to the severity of disease, *e.g.* a clinical grading of angina is from 0 (none) to 4 (angina at rest), *or*
   (iii) a quantitative measurement, *e.g.* forced expiratory volume or blood pressure.

4   Measures of Exposure

Potential environmental hazards vary enormously in their nature and complexity. They can be local 'hot spots', *e.g.* a particular lead smelter or nuclear installation, or general issues of low level exposure in the community, *e.g.* lead in the environment. Also, one can be concerned with a specific chemical, *e.g.* cadmium, asbestos, or a more complex potential hazard, *e.g.* sewage in water supplies. Hence, the degree of specificity of the measures of exposure in an epidemiological study depends on the type of hazard being investigated.

In studies of individual subjects, the ideal circumstance is to have a direct measure of individual intake or *body burden* of the environmental contaminant. However, even if measures of body burden do exist one still needs cautious assessment as to how appropriate they are in relationship to measures of ill health. For instance, one can measure lead in blood quite easily, but it only reflects relatively short-term exposure to environmental lead over the past month or two. In studies relating lead to blood pressure, blood lead may or may not be a suitable measure depending on whether the potential hypertensive effect of lead is hypothesized to be a short-term acute response or a longer term chronic effect of prolonged exposure. In studies of neuropsychological effects in children (say aged 6 or more) it has been thought that measuring lead concentration in shed milk teeth provides a more meaningful measure of long-term cumulative exposure, whereas in longitudinal studies of birth cohorts repeat measures of blood lead over time are satisfactory. Another example concerns the use of urinary cadmium excretion as a more appropriate measure than blood cadmium, since the former reflects the tendency for long-term cadmium intake to accumulate in the kidneys, but the latter reflects only very recent exposure.

The general principle is that an appropriate measure of body burden should be based on both a physiological understanding of the chemical in the body, and a realistic appraisal of the likely biological hypothesis relating the chemical to potential ill health. Unfortunately, for many chemicals (*e.g.* aluminium and asbestos) there is no readily available measure of body burden, in which case one has to resort to more 'environment based' measures which should be associated with actual intake. For instance, in occupational studies of asbestos exposure, one can classify jobs into various degrees of 'dustiness' (*e.g.* by use of on-site measurements of airborne asbestos).

A particular problem arises in studying individual exposure to chemicals in water supplies. For instance, it is hypothesized that an elevated nitrate intake may increase one's risk of stomach cancer, and that high water nitrate concentrations are thus undesirable. However, measuring individual consumptions of nitrates in water is practically infeasible. Instead, like many other issues concerning chemicals in water, one obtains measures of water nitrate concentration in the source of supply which therefore relates to 'whole community' rather than individual exposure. Such an indirect 'community' measure may provide a very poor guide to individual exposure, since people's use of their local water supply will vary enormously.

Another issue, particularly in studies of environment and cancer, concerns the time lag between exposure and development of disease, which may realistically be several decades. Thus, current measures of chemical exposure may be irrelevant in studies of cancer aetiology, except in so far as they may reflect past exposure. For instance, in a geographic study of reusage of water supplies and cancer mortality in the London area current measures of extent of reused sewage effluent in each community were related to community mortality rates of certain cancers, on the basis that broadly the same community differences in water reusage had existed for decades.

Overall, one key aspect in evaluating epidemiological evidence regarding chemical hazards is to critically assess the appropriateness of the exposure measures that have been used.

5   Confounding Factors

In all observational epidemiology, it is important to take account of confounding factors: *i.e.* other circumstances in the individuals or communities under study that may relate to both disease outcome and the environmental exposure. For instance, in the above-mentioned water reusage study, communities with high stomach cancer mortality tended to be the more socio-economically deprived (a finding compatible with the well-known social class gradient in many cancers). Areas of higher water reusage were poorer socio-economically on average, and hence a simple statistical analysis demonstrated an artificially strong association between stomach cancer and water reusage. Subsequent statistical adjustment for area social differences, using a multiple regression technique, greatly reduced this apparent association.

The identification of relevant confounding factors involves:

(a) judgements based on prior knowledge of the determinants and demographic associates of the disease under study, and

(b) *post hoc* statistical analyses to see which factors are related to disease in the sample studied. A distinction needs to be made between a true confounder and a factor that is directly in the causal pathway between hazard and disease. For instance, young children with pica (*i.e.* who ingest non-food items) are more prone to a high intake of lead, so that to adjust for pica (yes or no) in analysis of a cross-sectional survey might artificially reduce a true lead effect.

In evaluating the likely validity of an observed exposure/disease association one issue to consider is the possible relevance of unmeasured confounders. For instance, any study relating adult respiratory disease to a chemical hazard that fails to record and allow for the pattern of cigarette smoking has to be viewed with skepticism.

There may also exist synergism between an established aetiological factor and the environmental hazard. For instance, occupational studies of asbestos exposure have shown that the absolute increase in risk of lung cancer attributable to high asbestos exposure is much greater in smokers than non-smokers.

6   Avoidance of Bias and Measurement Error

The conduct of epidemiological studies requires particular care to avoid introducing bias, *i.e.* aspects of the observational study's procedure that artifactually either exaggerate or diminish the true environment/health relationship. Selection of subjects or geographic areas for study should aim to achieve a fair representation, possibly using some form of random sampling in studies of individuals. For instance, one study of heavy metals exposure in children recruited subjects by newspaper invitation, a technique that could not be reliably expected to represent community exposure.

Also, measurement techniques must be carried out in an unbiased manner. For instance, a study of lead and neuropsychological performance in children selected two groups of children with high and low tooth lead concentrations, but undertook neuropsychological testing on all the 'low lead' children first, thereby risking the possibility that any change in testing procedure over time could bias the comparison. Thus, individual measurements of exposure and disease outcome should be made independently (*i.e.* blind of one another) to avoid knowledge of the one having an influence on recording of the other.

Even if one's methods avoid biased measurement, it is important to consider the impact that unsystematic measurement errors can have on the observed association. Here we use the term 'error' to mean some, usually inevitable, random variation in observed measurement around the true (but unknown) value. This can be attributed to such features as genuine biological variation within subject over time, or variation in measurement technique (*e.g.* laboratory error).

Measurement error can exist for disease outcome, exposure, and confounder, the impact on the results being different in each case.

Errors in Disease Outcome

Use of just a single reading of systolic blood pressure, for example, will give a relatively poor guide to an individual's true underlying average systolic blood pressure. This will not bias the magnitude of association, *e.g.* on average the slope of the regression line for blood pressure regressed on blood lead (say) will not change because of such error. However, this random error in the dependent variable will affect statistical power, *i.e.* one will need a larger study to achieve statistical significance for a true exposure/disease effect. Use of averages of repeat readings on more than one occasion can help to reduce such variation in quantitative health measures.

Errors in Exposure Measurement

Lead concentration in a single incisor tooth shed at age six, for example, cannot perfectly reflect the child's true cumulative exposure to lead. Such error in the independent exposure variable will again result in loss of statistical power. However, it also reduces the observed magnitude of association, *e.g.* the slope of the repression line for (say) child IQ regressed on tooth lead will be less steep than the slope of child IQ regressed on the true but unknown lead exposure. Thus, imperfect measurement of exposure has an inevitable diluting effect on the observed association. In some environmental studies this issue is sufficiently important to question whether the study has any realistic chance of finding an association even if a real effect were present.

Errors in Measuring a Confounder

In a study of tooth lead and child IQ, for example, the most important confounding factor, parental IQ, was only obtained for the mother and then using only part of the full-scale adult IQ test. Such imprecise measurement of a confounder will have a similar consequence to not including the confounder at all. That is, only partial adjustment for a confounder, because of measurement error, will help to preserve an exaggerated association, the extent of exaggeration depending on the strength of the confounder's association with both dependent (health) and independent (environmental) variables.

The situation becomes particularly complex when both exposure and confounder are subject to measurement error. For instance, in the above example both tooth lead and parental IQ have sizeable errors. These errors compete simultaneously to dilute and accentuate the tooth lead–child IQ association so that an observed weak (non-significant) association after adjustment for mother's IQ is difficult to interpret.

7 Size of Study

The ability of an epidemiological study to demonstrate statistically that an environmental health problem does indeed exist is heavily dependent on the study's size. In discussing such statistical power it is useful to draw a distinction between studies of individuals and geographic studies of communities.

Studies of Individuals

This is not the place for a technical description of statistical power calculations, but the general principles can be made with examples. Cross-sectional surveys and case-control studies relating heavy metals, lead and cadmium, to adult blood pressure have mostly been carried out on fewer than 200 subjects. Since blood pressure varies markedly between subjects for reasons unrelated to environment (*e.g.* obesity, diet, heredity) such small studies are unlikely to detect an effect of lead (or cadmium) on blood pressure unless it is very marked. In occupational studies of high level exposure this might be plausible, but in studies of low level exposure in the general community one should realistically be looking for relatively weak associations. On this basis, two cross-sectional studies in Britain and the United States each based on several thousand subjects have demonstrated a very weak but highly significant association, estimated as around a 1 or 2 mmHg increase in systolic blood pressure for a doubling of blood lead. A causal relationship cannot be readily inferred (see below) but at least such large studies have been able to document the magnitude of statistical association with reasonable precision.

A more subtle problem with small studies is reporting bias, *i.e.* small studies that show a marked (exaggerated) statistical association between environmental exposure and ill health are more likely to be published (and receive widespread attention) than small studies that show no association. Thus, particularly in a field of investigation that suffers from studies being too small, it is important to evaluate the totality of evidence (some of which may be unpublished) rather than rely on the selective highlights of major journals which are prone to an exaggeratedly hazardous view of a potential environmental problem.

Another issue in considering study size is the rarity of the disease. For instance, a study of mortality in British armed forces personnel exposed to nuclear weapons testing followed over 20 000 subjects for over 20 years. The main findings from this massive study (with an equivalent-sized matched control group) were excesses of 16 leukaemia deaths and 6 myeloma deaths. Because of the rarity of these two causes of death, the study was actually quite small in terms of statistical power, and there remained some controversy over whether the excesses were genuinely causal.

Geographic Studies

In assessing the size of geographic studies, it is important to recognize that the units for statistical analysis are the geographic areas rather than individuals. Thus, a geographic study may appear very large in terms of the total population in the areas studied, but if the number of geographic units is small then the study may in a statistical sense be too small. That is, one requires adequate replication of geographic areas in order to demonstrate a convincing statistical association.

All too many geographic environmental studies have been based on just two areas with differing exposures. For instance, a Danish study compared stomach cancer mortality rates in two areas with high and low nitrate levels in water supplies. Such studies are inevitably inconclusive since in general area differences in disease are bound to exist for reasons other than the environmental factor of interest. In planning a geographic study one should aim for a substantial number of areas of differing environmental exposure. Areas should also be of sufficient population size and observation period to give precise estimates of disease rates, but should be sufficiently small to have homogenous exposure within each area. With a sufficient number of areas, one can then account for differences in confounding factors such as socio-economic differences.

8   Statistical Analysis and Reporting of Findings

Excessive reliance on tests of statistical significance can bedevil research in environmental epidemiology. The worst oversimplication is to conclude that the achievement of statistical significance demonstrates that the environmental factor causes ill health, and conversely that non-significance demonstrates that no such ill effect exists. The scope for *Type I* (*false positive*) and *Type II* (*false negative*) errors is considerable and needs careful assessment in any study.

First, Type I errors can arise due to the play of chance. That is, 1 in every 20 significance tests can expect to achieve $P < 0.05$ even if there is no real association present. Such chance considerations become particularly important if a study contains many significance tests relating to different measures of outcome and/or exposure, or if there are many similar studies.

Methodological flaws in study design may also give rise to Type I errors. It is a familiar theme to state that statistical analysis cannot rescue a poorly designed study, but with the high public and political profile that exists for many environmental issues, there continues to be a danger that statistically significant findings will still be interpreted as firm evidence of causality regardless of study quality.

However, even if a study has been conducted to the highest standards, with careful allowance for confounding factors and other potential biases, one still needs cautious interpretation of a highly significant finding. For instance, a Scottish study of over 400 6-year-old children found a highly significant inverse association between blood lead and child IQ ($P < 0.001$) even after allowance for various parental and social factors. The study paid careful attention to all methodological details in its design, conduct and analysis, but still one should be wary of jumping to the conclusion that moderate elevations in body burden of lead cause a deficit in child IQ. Other possible issues to consider are: (a) reverse causality, *i.e.* the possibility that children of lower IQ may adopt a behaviour pattern (*e.g.* more 'dirty play') which increases their lead intake, (b) selection bias, *i.e.* the unavoidable exclusion of children (and parents) who declined to participate in the study had distorted the association in some way, (c) residual confounding, *i.e.* it is difficult to measure all non-lead influences on child IQ so that part of the lead–IQ association might still be attributable to other factors, (d) blood lead measures only short-term exposure so that the observed association may not reflect the true effect of elevated lead exposure that would have occurred at younger ages. Thus, the cautious epidemiological scientist will need to assess the plausibility of these various alternative issues, in order to assess the validity of the causal hypothesis of prime interest.

This restraint in extrapolating from 'statistical association' to 'causal relationship' is an integral part of the epidemiological method, which principally applies because studies are observational rather than experimental.

It is important to recognize that hypothesis testing (*i.e.* use of *P* values) is only a limited aspect of statistical analysis in environmental studies. Indeed in *geographic studies* it is questionable whether hypothesis testing has any valid basis at all. After all, calculation of a *P* value is based on the concept that the units (*e.g.* people, geographic areas) are a random (representative) sample from some real or hypothetical population of such units. In geographic studies to hypothesize that the areas studied are a random sample requires some stretching of the imagination. Also, as mentioned in Section 2, geographic areas are likely to differ in ways that are (a) related to disease outcome, (b) not readily measured, and (c) not related to the hazard of interest. Hence, geographic (ecological) studies are usually a weak epidemiological approach not readily amenable to causal interpretation, as summarized by the well-known term 'ecological fallacy'. Thus, geographic studies are often best thought of as *descriptive* rather than analytical.

For instance, a recent study relating aluminium in water supplies to incidence of Alzheimer's disease used quite complex statistical analyses to reach a statistically significant association whereby subjects living in areas with low water aluminium concentrations had a lower incidence of Alzheimer's disease. However, such formal statistical inference for individual data is questionable, since both disease incidence and aluminium concentrations were determined not for individuals but for a number of geographic areas and their populations. That is, the unit of analysis was geographic, not individual, and a more geographically oriented display of data may have been more appropriate.

In environmental studies, it is important to *estimate the magnitude* of statistical association rather than merely quote a level of statistical significance. Also, use of confidence intervals is helpful in conveying the degree of statistical uncertainty inherent in any point estimate of association. For instance, a British study of cardiovascular mortality and water hardness in 234 urban areas demonstrated a statistically significant inverse association after allowance for socio-economic and climatic factors (but note the questionable merit of such a geographic P value). This study incorporated allowance for both measured co-variates and spatial autocorrelation, *i.e.* the fact that adjacent areas tend to have similar mortality rates for reasons that cannot be measured. The consequent estimate of the water hardness/cardiovascular mortality association was summarized as follows: areas of moderately hard water (3.0 mmol 1^{-1}) had a 7.8% reduction in cardiovascular mortality (95% confidence interval 3.7% to 12.0%) compared with areas of soft water (0.5 mmol 1^{-1}). This emphasis on estimation was helpful in conveying the fact that the apparently strong water hardness association with cardiovascular mortality in an initial straightforward univariate analysis was greatly reduced after allowance for other factors. Given that other personal risk factors (*e.g.* cigarette smoking) have over a 200% increase in risk of cardiovascular death, one felt able to conclude that the magnitude of association (even if causal) linking soft water to cardiovascular disease was relatively weak.

In environmental studies reporting negative findings (*i.e.* no statistical evidence of association between environmental factor and disease), one needs to beware of a possible Type II error (false negative conclusion). Type II errors are most likely to occur in small studies that lack statistical power, so that in assessing the validity of a negative study one should consider whether it was of adequate size. One indication of this is from confidence intervals, which in a small study will tend to be very wide, demonstrating that an apparently negative study could be compatible with a truly positive association.

9   Policy Decisions from Epidemiological Evidence

Sections 2 to 8 above have illustrated many of the fundamental issues to consider in the design, conduct, analysis, and interpretation of studies in environmental epidemiology. Overall, one must conclude that such observational study of potential environmental hazards in human populations is fraught with difficulties of interpretation. In so many areas of investigation this will make it extremely difficult to come to a definitive conclusion over whether a particular chemical exposure is or is not hazardous to human health.

The situation can be helped by having several different studies of the same environmental issue. It is often unwise to rely on a single study, with its particular sample of subjects and potential biases, so that environmental policy can be more broadly based if one has several *replicated studies*, of differing designs and populations. Of course, it is important that the totality of evidence be used rather than a potentially biased selection of the more publicized studies. It could be that one of the more important developments in the future of environmental epidemiology will be to formalize the manner in which one evaluates evidence from a collection of studies, for instance in using meta-analysis techniques.

The experienced and unbiased epidemiological scientist needs to assess in any given study or collection of studies what the inherent limitations are and to what extent they weaken the ability to place a causal interpretation on the observed associations, or alternatively to conclude 'no effect' in the absence of an association. Observational epidemiology is at its strongest when an exposure is well measured in individuals, the disease is readily identified, there are no major confounders, and the aetiological relationship is strong. A notable example is the link between cigarette smoking and lung cancer, where few would deny that prospective observational studies helped to establish a true causal association. Unfortunately, in environmental epidemiology we are rarely faced with such clear-cut circumstances.

This leads to apparent conflict between epidemiological caution and the needs of environmental policy-makers to make clear decisions on potential chemical (or other) hazards. In conclusion, there are two particular aspects worthy of specific comments:

(a) Many environmental health issues of greatest concern relate to *low level exposure applied to large populations*. For instance, many contaminants of water supplies (whether natural or introduced by man) are at low concentrations but such supplies serve whole communities. Hence even if such low level exposure does cause a very small increased risk of ill health to the individual, the fact that it is then applied to a large population still makes it an issue of some importance to public health. On the whole, epidemiological investigations are unlikely to be of great value in determining the effect (if any) of such low level exposure, unless the disease rarely occurs otherwise. For instance, there is considerable interest in whether elevated levels of nitrates in water supplies could increase the risk of stomach cancer but it is very doubtful if observational epidemiology could ever detect such an association (if it exists). Stomach cancer is a relatively common neoplasm; there are other notable associates with risk (*e.g.* social class), and nitrate in water is only one component of nitrate intake. Hence, absence of epidemiological evidence cannot be used as a strong argument for not limiting nitrate concentrations in water supplies. This leads to the second issue.

(b) How can we make sensible *recommendations on upper limits of exposure*? It is very important that national and international organizations, *e.g.* the World Health Organization, the European Economic Community, the US Environmental Protection Agency, make clear recommendations on what constitute unacceptably high exposures to potential chemical hazards. This requires specification of upper limits of exposure as for instance exists for a wide range of chemicals in water supplies. There is often an implicit (possibly false) assumption that there exists a *threshold* of exposure, below which there is no risk of ill health. Unfortunately, the concept of such a threshold is liable to be ill-founded because one may expect exposure/disease relationships to be a continuously increasing gradient of risk. Also, since we have already mentioned that epidemiology is unsuited to detecting weak relationships at low level exposure it seems doubtful whether epidemiological evidence is of great value in determining what recommended upper limits should not be exceeded.

Thus, environmental epidemiology is of greatest value in evaluating those excesses in exposure that lead to a marked increase in risk (*e.g.* the occupational exposure to asbestos that existed in the past), but the epidemiological approach is inevitably less productive in assessing the hazard (or safety) of widespread exposure to low levels of chemical hazards. This means that environmental policy-makers may need to exercise a greater degree of prudence by setting limits of exposure that are lower than can be justified by epidemiological evidence.

Chapter 7

Integrated Pollution Control: Application of Principles to Establish BPEO and BATNEEC

By Stefan Carlyle

ENVIRONMENT AGENCY, FORMERLY HER MAJESTY'S
INSPECTORATE OF POLLUTION, ROMNEY HOUSE,
43 MARSHAM STREET, LONDON SW1P 3PY, UK

1 Introduction

This paper describes new procedures being introduced by Her Majesty's Inspectorate of Pollution (HMIP) to carry out environmental, economic and Best Practicable Environmental Option risk assessments for Integrated Pollution Control. These procedures are likely to be further developed and used by the Environment Agency, which took over HMIP's regulatory role in April 1996.

2 Integrated Pollution Control (IPC)

The statutory basis for IPC is provided in Part 1 of the Environmental Protection Act 1990. IPC requires that no prescribed process can be operated without a prior authorization from HMIP. In setting the conditions to be attached to an authorization, Section 7 of the Act places HMIP under a duty to ensure that certain objectives are met. The conditions should ensure that:

- the Best Available Techniques (both technology and operating practices) Not Entailing Excessive Cost (BATNEEC) are used to prevent or, if that is not practicable, to minimize the release of prescribed substances into the medium for which they are prescribed; and to render harmless both any prescribed substances that are released and any other substances that might cause harm.
- releases do not cause, or contribute to, the breach of any direction given by the Secretary of State to implement European Community or international obligations relating to environmental protection, or any statutory environmental quality standards or objectives, or other statutory limits or requirements.
- when a process is likely to involve releases into more than one environmental medium (which will probably be the case in many processes prescribed for

IPC), the Best Practicable Environmental Option (BPEO) is achieved, *i.e.*
the releases from the process are controlled through the use of BATNEEC
to give the least overall effect on the environment as a whole.

3   IPC Applications

In applying for an IPC authorization an operator should:

(1) provide full information on the selection of the primary process, particularly
    for a new plant;
(2) provide evidence that the requirement to use BATNEEC will be met;
(3) select a combination of primary process, pollution abatement techniques,
    and waste treatment and disposal, which constitutes the BPEO;
(4) provide a justification for the BPEO selected and for all likely releases.

4   Environmental Assessments for IPC

Definition of Harm

It is clear from the objectives given in Section 7 of the Environmental Protection
Act (1990) that a means for assessing harm is required by HMIP for use in making
regulatory judgements. Within the context of the Act 'harm' means:

- "harm to the health of living organisms or other interference with the
  ecological systems of which they form a part and, in the case of man, includes
  offence caused to any of his senses or harm to his property; and 'harmless'
  has a corresponding meaning".

However, the Act does not define the nature of the effects that may be considered
harmful or the level in the environment at which they may occur. Therefore, to
overcome these difficulties a practical approach to the assessment of harm has
been devised and is described in the following section.

The proposed method for assessing the level of harm caused by individual
releases is illustrated in Figure 1. The 'intolerable level' is defined by either a
statutory limit such as an Environmental Quality Standard (EQS) for releases to
air or an Environmental Quality Objective for those to water. In the absence of a
statutory limit for a substance, an interim, Environmental Assessment Level
(EAL), will be set by HMIP to enable comparisons to be made. The EAL would
provide guidance to operators and inspectors on the tolerability of a predicted
environmental concentration of a pollutant.

HMIP Guide Levels can be used in assigning priorities for control; for example
if a predicted maximum ground level concentration is high compared with the
EAL (more than 80%) the release might command priority in generating
alternative process options. Similarly, if an individual plant's contribution to the
ambient level of a substance is more than 2% then it might be assigned a higher

Intolerable

_____ EQS (EAL)

	Predicted	
Tolerable	Environmental	
(priority	Concentration (PEC) of	
for	release (including	
control)	background) or	
	Process	
	Contribution (PC)	

_____ PEC = 0.8 of EQS/EAL or

Broadly   PC = 0.02 of EQS/EAL
acceptable

Figure 1   *Basis for decisions on the acceptability of releases*

priority for control.

The threshold value used to determine whether or not a release is significant is set at a level where we are confident that the effect on the environment is negligible. An example might be a factor of 100 less than the EQS (or EAL). Again a threshold value is being developed by HMIP as a guide to operators and site inspectors on the significance of releases where detailed environmental assessment would not be required.

Definition of the BPEO

Where a process involves the release of substances to more than one environmental medium the inspector will need to determine whether the proposed operation represents the BPEO for the pollutants concerned. However, since the Environmental Protection Act 1990 does not define the BPEO, it is proposed to adopt, with minor modification, the definition provided by the Royal Commission on Environmental Pollution in their 12th Report (1988) as follows:

- The BPEO can be considered as "the option which for a given set of objectives, provides the most benefit or least damage to the environmental as a whole, at acceptable cost, in the long term as well as the short term, *as a result of releases of substances from a prescribed process*".

The modification, highlighted by italics, has been added to bring the definition explicitly within the context of Section 7(7) of the Environmental Protection Act which limits the scope of the BPEO to consideration of substances released by the process.

The approach outlined above to assess the tolerability of releases can be used to assess the BPEO. Where tolerable significant releases occur (see Figure 1) to more than one environmental medium we can compare process/abatement options in an overall environmental (BPEO) context. Each significant release can be normalized by expressing the predicted environmental concentration as a percentage (or proportion) of the EQS (or Regulatory Assessment Level). The proportion of the EQS/EAL utilized by each significant release can be summed to create a tolerability quotient (TQ) for each medium, which is then used to derive an integrated environmental index (IEI) for a specific process/abatement option. This index could be used as the main basis on which to establish the best option from an environmental protection viewpoint. Additional environmental factors would need to be taken into account in ranking alternative process options (see below).

Site-Specific BATNEEC

The BPEO ranking derived from the above assessment would be combined with the economic or NEEC aspects to determine the site-specific BATNEEC option. A prescribed process should broadly meet the environmental performance of processes identified in Chief Inspectors Guidance Notes, either those actually described in the relevant note, or one which achieves an equivalent or better level of environmental protection. An operator should indicate the range of process/ abatement options considered in deciding on the eventual choice. If an option is proposed that clearly falls below the best environmental option then an operator would need to justify the choice in terms of cost effectiveness, *i.e.* by comparing the cost and environmental implications of the preferred and any discounted options. It will be for HMIP's site inspector to assess whether or not sufficient justification has been provided, perhaps by carrying out a separate assessment.

5 Environmental Assessment Procedure

Having established the main principles, it is necessary to develop a robust regulatory procedure. This should minimize the effort required by operators in preparing an application and by inspectors in assessing and confirming BATNEEC. It should also provide an audit trail documenting how the key decisions are reached. The stages in the procedure are as follows.

Stage 1: Preliminary Environmental Assessment

Step 1: Identify Pollutants Released.   For a specific (base case) process/abatement option identify the unavoidable releases of prescribed (and other potentially harmful) substances. Establish the release rates under start up, normal, and abnormal operation.

Step 2: Significance Test.   A significance test should be carried out to eliminate trivial releases from further consideration in the assessment. This will be based on advice provided by HMIP on levels (mass and/or concentration) or substances in releases that might be regarded as insignificant.

Step 3: Comparison with Chief Inspectors Guidance Notes (CIGNs).   CIGNs provide an indication of best practice for particular prescribed processes and the process in question should achieve the levels of environmental protection advocated in the relevant CIGN.

Step 4: Chimney Height Determination. Chimney height determination ensures that releases are rendered harmless so far as short-term maximum ground-level concentrations are concerned.

Step 5: Compute Predicted Environmental Concentrations (PEC). Use basic dispersion models to predict the maximum environmental concentration of each release. Add to this the actual or estimated background concentration of the pollutant at that point.

Compare the PEC for each significant release with any statutory limit (EQS), or where this is absent with the relevant EAL. (Those options resulting in a breach of the statutory limit would be automatically discounted.) Where a release results in a particularly high PEC compared with the EQS or EAL (typically greater than 80%) the Inspectorate might wish to examine this in more detail to reduce any uncertainties and to examine alternative process/abatement options.

Stage 2: Process Option Generation

The information gathered in Stage 1 about the environmental risks posed by the basic process can be used together with other relevant factors to generate alternative process options. Based on Stage 1 relevant environmental factors might include:

(1) reducing releases of substances that might lead to a break in an EQS;
(2) reducing releases where the PEC is high compared with the EAL, *e.g.* more than 80% of an EAL;
(3) reducing releases where the plant contribution would be more than 2% of an EAL;
(4) reducing releases to ensure that sensitive ecosystems are not harmed.

Other factors might also be relevant to selecting options for BPEO assessment including:

(1) health and safety
(2) physical space
(3) loss of amenity
(4) energy consumption

Stage 3: BPEO Assessment

Step 1: For Significant Releases Establish the Normalized Tolerability Quotient and Calculate the IEI. The PECs for significant release should be confirmed using more comprehensive predictive models. Releases are normalized as a percentage of the statutory limit (or interim EAL) to give the TQ. TQs for significant releases to all media are summed to give the BPEO index. The derivation of an IEI can be summarized as follows.

(1) For a tolerable single release

$$TQ = \frac{PEC}{\text{Statutory Limit}}$$

(2) TQ for all tolerable significant releases to an environmental medium

$$TQ^{(AIR)} = TQ_a + TQ_b \ldots + TQ_i \qquad \text{(for } a + b \ldots i \text{ releases)}$$

(3) BPEO Index

$$BPEO_{(index)} = TQ^{(AIR)} + TQ^{(WATER)} + TQ^{(LAND)}$$

(4) BPEO from a range of environmentally tolerable options.

$$BPEO = \text{Lowest BPEO index}$$

Step 2: Assessment of Other Environmental Factors.   Other relevant factors should be assessed to check the validity of the ranking of options based on the IEI. These might include short-term effects, global warming effects, potential to generate ozone, waste arisings, *etc.* These would need to be taken into account by carrying out some kind of multi-attribute analysis to rank the options in terms of environmental effects.

6   Short-term Effects

Different process options may lead to variations in the pattern of releases; for example, a process operated intermittently may give lower annual concentrations compared with one run continuously, but an increased frequency of short-term peaks may be the result. The assessment of short-term releases should therefore be an integral part of the environmental assessment under IPC for both new and existing processes.

7   Assessment of Global Warming Potential (GWP) of Releases

The release of carbon dioxide, water vapour, chlorofluorocarbons, methane, and nitrous oxide may contribute to global warming. In addition, tropospheric ozone (*i.e.* in the bottom 8–16 km of the atmosphere) also acts as a greenhouse gas, but the magnitude of this effect is still under investigation. It is important that the release of these gases is minimized wherever possible. Due to the nature of the effects arising from these pollutants it is not possible to incorporate them directly in the Environmental Index. Instead, process options should be ranked according to their potential to contribute to radiative forcing (global warming) expressed in carbon dioxide equivalents.

8 Assessment of the Potential for Ozone Generation

Ozone is a highly reactive pollutant that may exert a number of damaging effects on human health, vegetation, and materials. The production of ozone in the troposphere (*i.e.* the bottom 8–13 km of the atmosphere) involves the action of sunlight on hydrocarbons, usually referred to as volatile organic compounds (VOCs), and oxides of nitrogen (NO_x). The availability of NO_x downwind of a source controls the spatial extent of the area within which raised ozone concentrations may be generated.

There is a large variation between the importance of individual VOCs in their potential for ozone generation, depending on their reactivity with hydroxyl (OH) radicals and the subsequent production of peroxy (RO_2) radicals. In order to assess the relative effect of different hydrocarbons in the episodic production of ozone and provide a basis for their control, the UNECE VOC convention (1991) has proposed the concept of the Photochemical Ozone Creation Potential (POCP). The POCP is defined as the change in photochemical ozone production due to a change in emission of that particular VOC. The POCP may be determined by photochemical model calculations or by laboratory experiments.

Estimated individual POCP values will vary both temporally and spatially depending on the VOC composition of the modelled air parcel, the assumed meteorological conditions, and NO_x concentrations. However, although there is considerable uncertainty over individual POCP values the approach can be used to classify VOC species according to their importance in ozone production and average values assigned to each class of compound. It can be used, for example, to rank process options according to the overall POCP of the option.

9 Assessment of Waste Arisings

Many processes will generate quantities of solid or liquid waste, *i.e.* material that is not released to air or water. These wastes may be treated or disposed of on-site or removed from the plant for treatment or disposal elsewhere.

Incorporating such waste arisings into the calculation of an IEI is less straightforward than is the case with releases to air or water. This is because their environmental effects are likely to depend on a range of more complex interactions. A separate means of comparing this aspect of different process options is required, and this can be achieved by assessing each waste arising on the basis of quantity and relative hazard potential. The relative hazard potential will be determined by its physical, chemical, and biological characteristics.

A number of hazard assessment schemes can be found in the literature. The one proposed in this document is based on that developed by the UK Government–Industry Working Group on Priority Setting and Risk Assessment (1991). The scheme is based on a number of parameters:

(1) toxicity (to mammals and aquatic organisms);
(2) potential for bioaccumulation;
(3) degradation (in soil/water);
(4) other physical characteristics such as solubility, adsorption potential, and volatility.

Each of these parameters is scored and a total score is obtained by multiplying the individual scores. The score obtained represents the potential hazard for a unit quantity of the substance concerned. The 'unit hazard' can then be weighed by the quantity generated to obtain a final score for the substance concerned. The final scores for all the components of a waste can then be summed to give an overall hazard score for the process being considered. Different process options producing different wastes can then be ranked according to their overall hazard scores.

10   Other Environmental Factors

The above factors are neither exhaustive nor exclusive. Other factors that may be relevant at a particular site include: odours, visible plumes from chimneys, releases of dioxins or furans, releases of acid gases where critical loads might be exceeded, non-routine releases, *etc.* However, not all of these factors will be relevant to any one site and judgement will need to be used in deciding which factors are relevant, in both preparing and evaluating an IPC application.

11   Stage 4: Determine Site-Specific BATNEEC Option

From the range of environmentally tolerable options generated under Stages 1 and 2, the next step is to select the site-specific BATNEEC option. If the choice has the lowest BPEO index (greatest environmental protection) then no comparative economic analysis is needed. If not, then the reason for rejecting the more environmentally beneficial options should be justified in terms of cost effectiveness.

No formal rule for prescribing the choice of BATNEEC can be prescribed. (The operator must reach his own view as to which of the potential process/abatement options is the site-specific BATNEEC.) However the application of cost-effectiveness analysis assists the operator by reducing the number of options. This can be achieved in two ways.

(1) Options that are cost-effective can be identified. A cost-effective BATNEEC process/abatement option is defined as the one that achieves a given level of pollution control at least cost;

(2) The costs of achieving more stringent levels of pollution control can be shown. This information will illustrate any significant 'break point' beyond which reductions in pollution can only be achieved at greater incremental cost.

When identifying the least cost option, it is important to note that the cost-effectiveness result may be sensitive to the context in which an individual abatement technique is applied. For example, plant is designed for a specified range of operating conditions, and to operate beyond that range or under different conditions may entail additional operating and/or capital costs.

Two main techniques can be used to identify cost-effective treatment or abatement options:

(1) Annualized costs of different levels of pollution control can be plotted on a diagram, as illustrated in Figure 2; or
(2) Incremental cost-effectiveness can be presented arithmetically to compare the costs and abatement of a process option with those of the next more environmentally harmful option. This can be expressed more formally as:

$$\text{Incremental Cost} = \frac{\text{Cost Option 1} - \text{Cost Option 2}}{\text{Difference in pollution potential}}$$
$$\text{(Option 2} - \text{Option 1)}$$

Table 1 is an example of how the incremental costs of pollution control can be demonstrated. Other things being equal, in this case it might be concluded that the site-specific BATNEEC is option 3, because of the doubling of incremental cost to achieve any greater improvement.

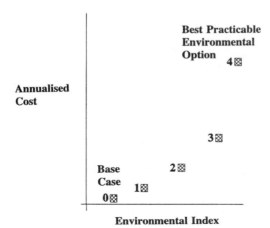

Figure 2   *Principles for establishing the site-specific BATNEEC option*

Table 1 *Illustration of the incremental cost of pollution control*

Option	Equivalent annual cost	Incremental improvement in BPEO Index	Incremental cost
Base case (uncontrolled)	£0	–	–
Option 1	£20000	1	£20000
Option 2	£50000	1	£30000
Option 3	£100000	1	£50000
Option 4	£200000	1	£100000

12   Judgement and Flexibility in Determining the BPEO Option

In identifying a 'break-point' where the costs of further reductions in pollution potential start to rise significantly, it is important to have a quantitative measure of the environmental consequences of the releases involved. Where the choice of the preferred option has been based largely on the long-term consequences of the releases as indicated by the Environmental Index, then this may provide a convenient measure of the overall pollution potential of a process (Figure 2). When the preferred option has been selected on the basis of other environmental factors, it may be more appropriate to use different measures of pollution potential such as the quantity of individual substances removed or the GWP of different abatement options. The important point is that the procedure is designed to set out the rationale for selection of the BATNEEC option. It should not be used in a mechanistic way but flexibly according to the particular nature of the process and in particular the nature and quantity of substances released.

Bibliography

The Environmental Protection Act 1990, HMSO, London, 1990.
The Environmental Protection (Prescribed Processes and Substances) Regulations 1991 SI No 472, as amended by the Environmental Protection (Prescribed Processes and Substances) (Amendment) Regulations 1992, SI No 614, HMSO, London, 1992.
Integrated Pollution Control – A Practical Guide, Department of the Environment, Marsham Street, London, 1991.
Environmental, Economic and BPEO Assessment Principles for Integrated Pollution Control – A Consultation Document, HMIP, London, April 1994.
Royal Commission on Environmental Pollution, 12th Report, Best Practicable Environmental Option, HMSO, London, 1988.
UNECE VOC Convention, 1991.
Department of the Environment, UK Government – Industry Working Group on Priority Setting and Risk Assessment, DoE, London, 1991.

Chapter 8

Guideline Values for Contaminated Land: Underlying Risk Assessment Concepts

By Colin C. Ferguson[1] and Judith M. Denner[2]

[1]CENTRE FOR RESEARCH INTO THE BUILT ENVIRONMENT, NOTTINGHAM TRENT UNIVERSITY, BURTON STREET, NOTTINGHAM NG1 4BU, UK
[2]DEPARTMENT OF THE ENVIRONMENT, ROMSEY HOUSE, 43 MARSHAM STREET, LONDON SW1P 3PY, UK

1 Introduction

Risk analysis is sometimes seen simply as a scientific exercise to determine what the risks 'really are'. The reality is more complex. The use of risk analysis to place limits on contaminant concentrations in food, water, air, or soil cannot be divorced from social and economic considerations. Nor are the facts and the scientific methodologies that underpin standards and guidelines wholly divorced from these wider considerations. Jasanoff[1] observed that "the principles by which we organize the 'facts' of risk have to derive, at least in part, from underlying concerns of public policy and social justice: whom should we protect, against what harms, at what cost, and by foregoing what other opportunities?"

The main purpose of this chapter is to discuss how scientific and other factors interact in the derivation of standards or guidelines for contaminants in soil. There are two main reasons why such a discussion is important.

(1) Decisions that have implications for human health and environmental protection, but which might also impose substantial cost on industry, need to be made in an explicit way. This accords with the principles of public accountability and open government. Also, much of the unease about environmental risks "probably stems from the public's lack of trust in many of the individuals, industries and institutions responsible for risk analysis and risk management".[2] Openness, or lack of it, thus affects risk perception (and risk acceptability) which in turn can influence the standards or guidelines that are eventually set.
(2) There has been much concern, both in the UK and elsewhere, that soil guidelines are frequently misused. The misuse usually results from misunderstanding. Open discussion of the factors involved in setting new guidelines will assist in appropriate use of those guidelines and will thus improve the quality of environmental risk management.

This chapter does not attempt to define what is, or is not, an 'acceptable' risk. This can only be determined within the wider policy framework for managing environmental risks. The issues addressed in this chapter, however, are relevant to that decision-making process.

2   Hazard, Risk, and Safety

The first point to consider is the underlying meaning of the terms used in the process of risk assessment. Confusion over meaning often arises because many terms have both a formal technical meaning and a looser everyday meaning. *Hazard, risk,* and *safety* all carry shades of meaning that depend on culture and context.

A HAZARD can be considered to exist if a potential exists to cause harm or damage *by virtue of the properties of a substance and the circumstances in which it occurs.* A toxic chemical is thus not necessarily hazardous under this definition if the circumstances in which it occurs (suitable container in a locked laboratory cabinet) preclude the potential for harm. Pure water could be described as hazardous if the circumstances involve high temperature or pressures. But *hazard* is often used without qualification to indicate a potential for causing harm. In this context describing a substance as hazardous refers only to the *properties* of the substance.

RISK is used with several shades of meaning even in the technical literature. In environmental risk it is usually used to refer to the *probability* that a hazard will result in specified harm or damage to a specified receptor. In this sense it combines the concepts of frequency of occurrence of a hazardous event with the consequences of the occurrence. But for certain applications risk can also be defined as the probability that a particular adverse *event* occurs during a stated time period. This avoids the need to specify *particular* types of harm or damage and is similar to the way *risk* is used in everyday speech: to mean the chance of a bad outcome, without necessarily specifying the outcome in detail.

SAFETY can be described as freedom from risks of harm. In a world of hazards there can be no absolute safety and therefore safety is usually defined as freedom from *unacceptable* risks of harm. The acceptability of risks is very different in different contexts. What might be *accepted* as safe in rockclimbing would be considered extremely unsafe in a shopping trip.

There are no absolutes in environmental risk either. What may be *accepted* as safe depends on the circumstances, in particular the perceived costs and benefits.

3   Risk Analysis of Contaminated Sites

Human Health Risks

Sources of soil contamination include a wide range of industrial and waste disposal practices, both current and historic. Hazardous substances in soil obviously have the potential to cause harm or damage, but the chance of harm occurring (*i.e.* the risk) depends on a great many factors. The standard approach for analysing risks starts with a hazard assessment: identifying and assessing the various source-pathway-target possibilities and eliminating those that are not significant. For contaminants such as cadmium and mercury detailed source-pathway analysis to human targets is essential, although the relevant pathways will depend on the use proposed for a site. Copper and zinc, however, are not hazardous to human health at concentrations hundreds of times greater than those that are toxic to plants. Plants are therefore the critical target and a source-pathway-target analysis for human targets is unlikely to be necessary.

The next stage in risk assessment for human targets involves estimating the rate of intake of a contaminant from the various sources and via the various pathways, and evaluating the likely effects on health. Doing this on a site-specific basis should allow local factors to be taken into account, and lead to a more realistic estimation of the risk. But detailed site-specific risk analysis can be complex and expensive, especially for sites where several different targets and many different pathways need to be considered. It requires substantial expertise in a variety of scientific disciplines, and it can be difficult to obtain consensus on the most appropriate approach. An alternative approach is to make decisions on the suitability of soil for specific site uses on the basis of guideline values. Developed and used correctly, this provides a nationally consistent and simple framework for decision-making.

There are common elements between the two approaches. For example, the major concern for human health is usually the effects of chronic exposure to low levels of soil contamination. Thus site-specific analysis, say for a proposed residential use, involves predicting levels of exposure over many years (typically 70 years) from ingestion, inhalation, and skin contact. The analysis therefore relates to a *typical person* whose physical characteristics (body weight, respiration rate *etc.*) and activity patterns need to be assumed. But exactly the same assumptions are made in using modern risk-based methods to derive soil guidelines or standards. The only difference is that, in site-specific analysis, full account can be taken of site characteristics such as soil type, aspect and climate, particular behaviour patterns, background intake, *etc.* Soil type, for example, is likely to influence the outcome significantly. But soil guidelines can also take account of differences in soil type (*e.g.* Dutch target and intervention values[3] and current Department of the Environment research to revise and expand trigger values[4]) and be explicit in the way they take into account the other factors.

Ideally guideline values can be *integrated* with site-specific risk analysis to provide a consistent framework for assessment, which also ensures that local circumstances are taken into account. In an *integrated* approach, an action (trigger) value based on general assumptions could either lead to the need for remedial action, or could lead to a further evaluation of risks using site-specific assessment. The latter course of action does not necessarily imply the slow and costly procedure that site-specific assessment has often been in the past. As indicated above, it may mean no more than identifying those site conditions that are significantly different from the general conditions assumed in deriving the trigger value: and then deciding on whether those differences would make a significant difference to the outcome (*i.e.* long-term contaminant intake). Though not necessarily costly or time-consuming, this does require that the exposure assumptions and data values used in deriving trigger values are available for scrutiny so that the necessary comparisons with site-specific values can be made.

Groundwater and Surface Water Risks

It is more difficult to develop generic soil guidelines for groundwater protection. This is because most of the key variables (thickness and attenuating capacity of soil and bedrock, depth to water table, source protection zone status, *etc.*) are highly site-specific. Generic guidance serves a somewhat different role, for example to provide a rough indication of the conditions under which soil guidelines derived for direct human health protection might need to be adjusted downwards to protect groundwater quality, or to provide a set of worst-case assumptions to screen out sites (*e.g.* leachability, use of drinking water standards). But groundwater risks should always be assessed on a site-specific basis.

Similarly, transfer of contaminants from soil to surface waters is highly site-specific and depends on run-off volume, peak flow rate, soil erodibility, slope length and steepness, sorption capacity of the soil, vegetation cover type, and distance to receiving body. For many sites the major threat to surface water quality is likely to occur during site redevelopment, and will need to be assessed on a site-specific basis. There is little merit, therefore, in trying to derive general soil guidelines to protect surface waters. Site assessors need to be aware that guideline or trigger values, or other guidance, often serve a narrower purpose and cannot be expected to indicate, even roughly, whether soil contaminants might impact adversely on surface waters.

The Department of the Environment has recently published guidance on the assessment of the impact of contaminated land on the water environment.[5]

Other Targets

Human health and the water environment are not the only targets that could be affected by contamination on a site. The wider ecosystem, buildings, and other structures or materials may also be vulnerable. These are being considered in the Department's research programme (*e.g.* reference 6).

4   Toxicology: Criteria and Safety Conventions

Threshold Compounds

A key step in risk analysis is comparing estimated human intakes of soil contaminants with 'safe' levels determined by toxicologists. It is an established principle of toxicology that, for many substances, there is a dose level below which no adverse effects are observed. This is referred to as a 'dose threshold' and substances that display such a threshold are often called *threshold compounds*. Genotoxic carcinogens, however, are not believed to have such a threshold.

In theory, then, it is easier to decide on a safe dose for threshold substances, *i.e.*

determine the maximum dose at which no adverse effects occur, than for genotoxic carcinogens. But the information about dose thresholds is not derived from controlled toxicity experiments on human subjects. Such experiments are out of the question. If they were to be conducted, progressively lower doses would almost certainly be associated with successively less severe effects until, at the biological level, some reaction always occurs. There is thus no 'threshold'.[7] In addition, the effect of a dose depends on a number of factors, including absorption, distribution, biotransformation, retention or accumulation, and elimination: all these factors are important in the determination of thresholds, because they all affect the number of active molecules that eventually come into contact with the biological system in which the toxic reaction occurs. These considerations suggest that the adverse effect threshold for a toxic substance will depend on characteristics of the individual organism. Each member of a population in effect has its own threshold.

In practice these complexities are overcome by the convention of determining a No Observed Adverse Effect Level (NOAEL) from animal experiments, and then deriving a human Acceptable Daily Intake (ADI) by dividing the animal NOAEL by a safety factor. Traditionally the factor used is 100 representing two factors of 10, one to allow for interspecies variation (animal to human) and the other to allow for variations in sensitivity among humans. These factors date back to the 1950s when two scientists from the US Food and Drug Administration reviewed the rather scant literature on relative sensitivities of humans and test animals to chemicals used as food additives. Their conclusion at that time, that a safety factor of 100 was adequately protective, is still widely used not only for food additives but also now for other environmental media. So the traditional method of determining a human ADI is a 'convention', as it is not solely the result of detailed scientific investigation. Sometimes further safety factors are incorporated to allow for animal studies that are subchronic rather than chronic (once more, a factor of 10 has been chosen traditionally) or to compensate for statistical bias towards the absence-of-effect hypothesis.[8] Again these are no more than conventions.

There is no exactness about an ADI, or about any soil guideline or standard derived from it. It is therefore more appropriate to consider a soil guideline or trigger value as an inexact point within a broad zone for safety, rather than as a sharp dividing line between 'safe' and 'unsafe'.

Genotoxic Carcinogens

There is a special difficulty with genotoxic carcinogens, which are assumed to have no safe-dose threshold. The result of animal tests (cancer bioassays) is usually an increase in the incidence of tumours in the dose group relative to the control group. But the excess lifetime risk of cancer that can be detected and judged to be statistically significant depends on the number of animals used. In practice it is difficult to detect excess lifetime cancer risks of less than 5–10%. But for almost all carcinogens in the environment, human exposures (and the resulting doses) are orders of magnitude smaller than those used in animal studies. Thus estimating excess cancer risks from such low levels of exposure involves not only applying animal test results to humans but also extrapolating to cancer risks thousands of times smaller than those for which dose–response data are available.

Numerous mathematical models for low-dose extrapolation have been proposed, but most fall into three general types as shown in Figure 1. The *linear* model has its origins in research on radiation-induced cancers and on certain chemicals that seem to mimic radiation in the way that they increase cancer risk. *Sublinear* models are considered by some to provide a better description of the low-dose risk relationship because they take account of biological mechanisms that reduce the effectiveness of genotoxic agents at low doses. Others propose a *threshold* model on the basis that each organism can tolerate a certain amount of exposure to any toxic agent, including carcinogens. (The drawback in this case is that there is no practicable means of verifying, even approximately, the threshold dose.) Results will differ: a particular human dose from environmental exposure (say dose B, Figure 1) might be associated with no excess risk of cancer (threshold model) or with excess risks that differ by an order of magnitude depending on whether the linear or sublinear model is used.

Accepted levels for contaminants in the environment are often determined using the reverse procedure. A level of excess lifetime cancer risk (say 10^{-4} or 1 in 10 000, Figure 1) is set and then the maximum dose that would not exceed that risk is estimated. This procedure can be misunderstood and lead to the excess risk

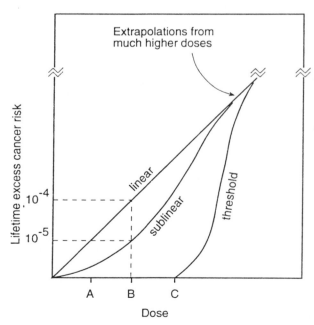

Figure 1   *Hypothetical dose–response curves for chemically induced cancer at low doses*

criterion (say, 10^{-4}) being treated as a *quantitative* statement about risk. This is not the case. The authors of pioneering work on extrapolating cancer bioassay data to low doses[9] did not suppose that their procedure could be used to predict the actual risk associated with a given low dose. Rather, they saw it as a method for identifying the dose that would be unlikely to create a risk greater than some operationally defined 'safe' level. From the outset, quantitative risk assessment for genotoxic carcinogens has been a 'convention', to express the concept of 'virtually safe'.

It is not valid for example to conclude from a target concentration of contamination based on a 10^{-6} risk criterion that if 55 million UK citizens were all to live on soil contaminated at the 10^{-6} guideline level then this would result in 55 excess deaths from cancer in the population. Equally, on the basis that 1 in 4 people in Britain die from cancer (a background rate of 250 000 in 1 million) it would be preposterous to claim that persons exposed to a '10^{-6}' concentration have a lifetime risk of cancer mortality of 250 001 in 1 million.

In fact it is not possible to predict how many excess cancers (if any) would result from lifetime exposure to such a concentration theoretically based on this 10^{-6} criterion. As noted by Dr Frank Young, the former commissioner of the US Food and Drug Administration (FDA), when the FDA uses the risk level of 1 in 10^6 "it is confident that the risk to humans is virtually non-existent, rather than 1 in 1 million exposed persons is expected to develop cancer".[10] The term *de minimis* risk is useful in conveying the concept of virtually non-existent risk. It is now common in the USA to refer to risk levels such as 10^{-6} as *theoretical* risks and to dissuade people from calculating excess cancer deaths. The US Environmental Protection Agency (EPA) advises that "these estimates represent an upper bound of the plausible risk and are not likely to underestimate the risk. The actual risk may be much lower, and in some cases, zero".

Of course, someone still has to choose whether 10^{-6} or some other number is the appropriate theoretical risk to express the concept of 'virtually safe'. Mantel and Bryan,[9] for example, chose the very small number of 10^{-8}, or 1 in 100 million, as their theoretical risk convention. But one cannot conclude that soil guidelines based on this risk convention would be 100 times safer than guidelines derived from a 10^{-6} risk convention. Similarly where the new Dutch intervention values for genotoxic carcinogens are based on a 10^{-4} theoretical risk convention it is not valid to say that these are 100 times less safe than guidelines derived from a 10^{-6} criterion. This is because *inter alia* the conservative assumptions used in *exposure* assessment (discussed later) can also be thought of as safety conventions.

A regulator's decision on an appropriate risk convention is not, therefore, made in isolation but with knowledge of the other safety conventions included, explicitly or implicitly, in the whole exposure and risk assessment process.

Occupational Epidemiology

There is a reluctance in the UK to place too much emphasis on derivation of guidelines from laboratory studies of rats and mice because of the difficulty in extrapolation from high-dose results in animal tests to effects in people at very much lower doses. As a general principle, the Department of Health Committee on Carcinogenicity prefers to put greater weight on studies of human populations, especially those exposed to chemicals in the workplace.[11,12]

Nevertheless this approach presents difficulties. Industrial workers may be exposed to higher concentrations of chemicals than the population at large, but usually do not include the more sensitive members of the population (children, the elderly, the sick, pregnant women). Extrapolation from relatively high doses in healthy young and middle-aged male workers to lower doses in the general population is rarely straightforward because of uncertainties over actual exposure levels, simultaneous exposure to other chemicals, differences in sensitivity, and different total exposure times. To overcome these difficulties safety factors are introduced, which again are no more than conventions chosen to provide reasonable confidence that certain very low levels of exposure are 'virtually safe'.*

5 Other Safety Conventions and Assumptions

As well as toxicological safety factors, another type of safety convention is implicit in many risk assessments, that is the use of conservative assumptions in *exposure* assessments. Keenan *et al.*[13] draw attention to a recent example from the USA of an analysis based on a hypothetical Maximally Exposed Individual (MEI):

"... at first review, the analysis seemed reasonable until one noted that the child ate about a teaspoon full of dirt each day, that his house was downwind of the stack, that he ate fish from the pond near the incinerator, his fish consumption was at the 95 percentile level, he drank contaminated water from the pond, he ate food grown primarily from the family garden, and he drank milk from a cow that had grazed on forage at the farm."

Such conservative exposure assumptions are increasingly being challenged. The exposure assessment approach favoured in the UK for deriving soil guidelines does not assume a hypothetical MEI. Instead it is based on using the Contaminated Land Exposure Assessment (CLEA) model to derive a realistic *distribution* of exposure levels among an exposed population appropriate for the

---

* For example in deriving an Air Quality Standard for benzene, the Expert Panel on Air Quality Standards[12] first reviewed studies of occupational exposure to benzene and concluded that the risk of cancer (in particular non-lymphocytic leukaemias) in workers exposed to 500 ppb benzene in air over a working lifetime would be too small to detect in any feasible study. They then divided this concentration by 10 to allow for the differences between a chronological lifetime and a working lifetime. Finally they introduced another safety factor of 10 ('analogous to factors used in regulatory toxicology for non-carcinogens') to protect sensitive subgroups in the general population. The resulting recommendation for an Air Quality Standard for benzene, 5 ppb as a running annual average, is considered by the Expert Panel to be a concentration "at which the risks are exceedingly small and unlikely to be detectable by any practicable method".

proposed end use.[4] This provides a more explicit and realistic framework for decision-making.

Further conventions are involved in the selection of sampling regimes and analytical techniques to determine the actual concentration of contaminants on a particular site. Sampling regimes carry their own underlying statistical assumptions about the representativeness of a sample. Analytical techniques are also rarely absolute, and may not truly reflect the particular 'hazard' represented by the contaminant for which guidelines are being set; for example there may be an additional factor of safety in measuring total concentrations of metals rather than their bioavailable concentrations.

6 Setting Soil Contaminant Guidelines: A Risk Management Philosophy

The framework of guidance or regulation in which risk management decisions are taken needs to balance the need to safeguard human health and the environment and the need to foster activities that will ultimately fund remedial works.[14] As Figure 2 illustrates, the cost of risk reduction can be highly non-linear. If soil guidelines are set at levels that are prohibitively costly to achieve, redevelopment of many old industrial sites will not go ahead. The costs of untackled dereliction might include loss of new job opportunities or new amenities, and reduced local property values and quality of life. Failure to

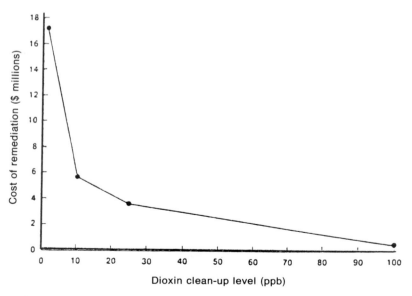

Figure 2   *Estimated cost of soil removal and incineration for various target soil clean-up levels at a site in Missouri, USA*
From reference 15

redevelop sites increases the pressures for greenfield sites to be converted to industrial or commercial use.

However, the risks to the public from contaminated land are involuntary and often not known or understood completely. Consequently the level of concern is typically much greater than might be expected from objective analysis of the risks. There is nothing irrational about this. Everyone expects a much higher level of protection against involuntary risks compared with risks from activities willingly entered into and bringing clearly identifiable benefits. The goal of risk management, then, is to select options that balance the costs of an action against protection from the real and *perceived* risks.

From the earlier discussion it can be seen that soil guidelines (*e.g.* action trigger values) should not be thought of as marking a sharp boundary between 'safe' and 'unsafe' concentrations in soil. It is more appropriate to think in terms of gradational zones from 'very safe' to 'not safe' with a diffuse boundary zone between them (Figure 3). Scientific risk analysis can aim to relate these zones to contaminant concentrations in soil. Choosing a *particular* value to act as a guideline is not, however, an entirely scientific decision. It is a risk management decision in which benefits and costs are also balanced.

7   The Use of Guideline Values

It is also necessary to establish the framework in which the guideline values would be used. Guidance and legislation in different countries are variously based on one or two guideline values for individual contaminants, sometimes combined with a third number representing background concentrations or

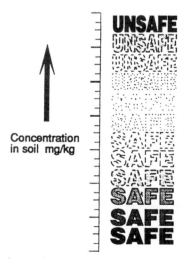

Figure 3   *Schematic illustration of 'safe' and 'unsafe' zones separated by a diffuse boundary zone*

analytical detection limits (*e.g.* Dutch target values). Soil clean-up targets may coincide with action levels or be set at lower levels.[15]

In the UK the present system is based on *threshold* and *action* trigger values.[16] Very few action values have been set and this has led to uncertainty over the level at which a regulatory body might impose requirements for investigation of a particular site. As part of the Government's research programme to confirm or revise existing trigger values, and extend the coverage to all the main contaminants in soil, the emphasis has therefore been on developing *action levels* for human health protection.

As with current Interdepartmental Committee on the Redevelopment of Contaminated Land (ICRCL) guidance, action levels would be set according to the proposed site use reflecting the Government's 'suitable for use' approach.[14] Action levels for human health protection are also expected to be related to soil type where this results in significantly different levels of exposure to soil contaminants.

Where an action level is exceeded there are two possible courses of action. The first is to undertake an appropriate *site-specific risk assessment* so that site characteristics that differ significantly from the generic assumptions implicit in setting the action level can be taken into account. This might lead to the conclusion that remedial action is not warranted. The second possible course of action is remedial work to reduce the concentration of soil contaminants or to reduce exposure to the contaminated soil. Action values derived from a generic basis are therefore used as a *guide* to appropriate and cost-effective action.

As an alternative to setting specific target levels for soil clean-up, risk reduction can be guided by the ALARP principle, that remedial action should be taken to reduce soil contaminant concentrations to levels that are *as low as reasonably practicable*. Reasonableness and practicability will depend on local circumstances, but in general the BATNEEC principle will be relevant (that the *best available techniques not entailing excessive cost* are used) as will consideration of the costs and benefits of various options.

Of course in some circumstances action may be needed where concentrations do not exceed action levels. As ICRCL Guidance on the assessment and redevelopment of contaminated land states,[17] at contaminant concentrations *below* action trigger levels, there is still

> "a need to consider whether the presence of the contaminant justifies taking remedial action for the proposed use of a site. If such considerations suggest that some action is justified, then it should be taken: the decision to do so is therefore based on informed judgement."

Action trigger values can, of course, be set more and more conservatively to avoid this situation, but this would have a trade-off with the cost implications for the increased number of sites that would be investigated with more stringent action values. The alternative, which is that positive confirmation that the guideline values are appropriate and 'safe' for the site, may be less onerous.

The Department's work on setting action values for the protection of human health includes studies to review intake values based on established exposure criteria, occupational health criteria, and background exposure in the UK. These will provide input to the CLEA model discussed earlier.[4]

8   Conclusion

Public confidence in environmental policy and risk management ultimately requires confidence in scientists, political decision-makers, and regulators. Ruckelshaus,[17] a former administrator of the US EPA, said:

> "To effectively manage the risk, we must seek new ways to involve the public in the decision-making process ... For this to happen, scientists must be willing to take a larger role in explaining the risks to the public, including the uncertainties inherent in any risk assessment. Shouldering this burden is the responsibility of all scientists, not just those with a particular policy end in mind."

Acknowledgement

This work is funded by the Department of the Environment but the views expressed are those of the authors and do not necessarily represent those of the Department.

References

1 S. Jasanoff, 'Bridging the two cultures of risk analysis', *Risk Anal.*, 1993, **13**, 123–129.

2 P. Slovic, 'Perceived risk, trust and democracy', *Risk Anal.*, 1993, **13**, 675–682.

3 R. Van den Berg, C. A. J. Denneman, and J. M. Roels, 'Risk assessment of contaminated soil: proposal for adjusted, toxicologically based Dutch soil clean-up criteria', in 'Contaminated Soil '93', eds. F. Arendt, G. J. Annokkee, R. Bosman, and W. J. van den Brink, Kluwer Academic, Dordrecht, 1993, pp. 349–364.

4 C. C. Ferguson and J. Denner, 'Soil guideline values in the UK: new risk-based approach', in 'Contaminated Soil '93', eds. F. Arendt, G. J. Annokkee, R. Bosman, and W. J. van den Brink, Kluwer Academic, Dordrecht, 1993, pp. 365–372.

5 Department of the Environment, 'A Framework for Assessing the Impact of Contaminated Land on Groundwater and Surface Water', CLR Report No. 1, Department of the Environment, London, 1994.

6 Building Research Establishment, 'Performance of Building Materials in Contaminated Land', BRE Report BR 255, 1994.

7 C. Cox, 'Threshold dose–response models in toxicology', *Biometrics*, 1987, **43**, 511–523.

8 J. A. Hoekstra and P. H. van Ewijk, 'Alternatives for the no-observed-effect level', *Environ. Toxicol. Chem.*, 1993, **12**, 187–194.

9 N. Mantel and W. Bryan, 'Safety testing of carcinogenic agents', *J. Natl. Cancer Inst.*, 1961, **27**, 455–470.

10 F. A. Young, 'Risk assessment: the convergence of science and law', *Regul. Toxicol. Pharmacol.*, 1987, **1**, 179.

11 Department of Health, 'Guidelines for the Evaluation of Chemicals for Carcinogenicity', Committee on Carcinogenicity of Chemicals in Food, Consumer Products and the

Environment, Report RHSS 42, HMSO, London, 1991.
12 Department of the Environment, 'Expert Panel on Air Quality Standards: Benzene', HMSO, London, 1994.
13 R. E. Keenan, E. R. Algeo, E. S. Ebert, and D. J. Paustenbach, 'Taking a risk assessment approach to RCRA corrective action', in Proc. of Developing Clean-up Standards for Contaminated Soil, Sediment and Groundwater: How Clean is Clean?, Water Environment Federation, Alexandria, 1993, pp. 225–275.
14 Department of the Environment/Welsh Office, 'Framework for Contaminated Land: Outcome of the Government's Policy Review and Conclusions from the Consultation Paper *Paying for Our Past*,' Department of the Environment, London, 1994.
15 D. J. Paustenbach, H. P. Shu, and F. J. Murray, 'A critical analysis of risk assessment of TCDD contaminated soil', *Regul. Toxicol. Pharmacol.*, 1986, **6**, 284–307.
16 W. J. F. Visser, 'Contaminated Land Policies in Some Industrialised Countries', Technische Commissie Bodembescherming, The Hague, 1993.
17 Interdepartmental Committee on the Redevelopment of Contaminated Land (ICRCL), 'Guidance on the Assessment and Redevelopment of Contaminated Land', ICRCL 59/83, 2nd edn, Department of the Environment, London, 1987.
18 W. D. Ruckelshaus, 'Science, risk and public policy', *Science*, 1984, **221**, 1026–1028.

Chapter 9

Contaminated Land and Water Quality Standards

By R. C. Harris and C. A. Thomas

ENVIRONMENTAL AGENCY, FORMERLY NATIONAL RIVERS
AUTHORITY, SEVERN-TRENT REGION, SAPPHIRE EAST,
550 STREETSBROOK ROAD, SOLIHULL B91 1QT, UK

1 Introduction

There are two main ways in which contaminants in polluted land can affect the environment. The first is by direct contact with the ecosystem and the second is from the migration of contaminants to groundwater and surface waters. This paper addresses the latter issue. Contaminated land and water quality issues have become more linked in recent years. It is now recognized that areas of past and present industrial activity have given rise to a legacy of contamination of the underlying soils and bedrock that has a potential to affect groundwater and surface water resources in the locality.[1,2] The potential depends upon many factors that relate to the characteristics of the contaminants, the underlying geology, and the use to which the land is put. These factors are complex and usually site-specific. Nevertheless there has been a call from industry for general standards to which land/soils/groundwater should be restored in order to prevent continuing pollution. This demand has increased with both the recognition of the role of the National Rivers Authority (NRA) in enforcing pollution prevention of the water environment (a role taken over by the Environment Agency in April 1996)[3] and the attitude of financial backers and insurers of developers and operators of contaminated sites in avoidance of residual liability.

2 Existing Guidance on Clean-up Standards

The UK guidelines for contaminated soils published by the Department of the Environment (DoE) on behalf of the Interdepartmental Committee on the Redevelopment of Contaminated Land (ICRCL) were developed as a result of a similar need.[4] The issue of a Circular to Local Authorities in 1977 raised awareness about the dangers of building on contaminated land. This resulted in the publication of guidelines or 'trigger values' in 1979, although there was an initial reluctance by the DoE to produce numbers for formal use.[5] These have been updated and extended, being presently contained in a document published in 1987.[6]

The ICRCL values are generic guidelines, intended to be used with professional judgement as guidance related to the end use of the redeveloped land. They are not comprehensive nor do they have any statutory backing. They also do not relate to the potential to affect water quality since the values are set for the total concentrations of each parameter in the solid phase. There are no corresponding guidelines for contaminated groundwater. The other commonly utilized UK guidance are the so-called 'Kelly standards' developed for the Greater London Council to aid in the assessment of the degree of contamination present at sites undergoing redevelopment.[7] These also bear no relevance to the water environment but are still widely used. Lately, Waste Regulatory Authorities have started to adopt them for the assessment of materials suitable for landfilling at unlined 'inert' sites. This is an example of the inappropriate adoption of a set of numbers that were developed for use in one context but have been taken up for another because of the absence of alternatives.

In recent years interest has been shown in the Dutch ABC List.[8] The ABC values represent respectively: background, not contaminated; contaminated, investigate further; contaminated, initiate clean-up. They relate to both soils and groundwater and therefore have been widely used in assessing the quality of leachate derived from leaching solid contaminated soil samples. Re-evaluation of the guidelines over the last few years has resulted in a more risk-based approach being proposed for future use in the Netherlands.[9]

All three of these sets of guidelines that have achieved widespread use in the UK were published with recommendations that the figures be used with caution and that other site-specific information is used to assess specific situations. Such recommendations, however, often go unheeded with the result that preparatory work for remediation can be well advanced on a site before an expert view on the actual risks is taken. Often site investigations have little regard for the hydrogeological/hydrological setting. There are many examples where the NRA, as statutory consultee of Local Planning Authorities, has required more extensive clean-up or remedial engineering to be undertaken because of particularly sensitive situations that were not understood at the outset. The converse is also true. Some sites present little threat to the water environment and the clean-up measures needed may be minimal.

Therefore working to a set of numbers regardless of the site-specific situation could not only lead to continuing residual liability, but also involve a developer in excessive costs. It presents even more of a problem when the numbers do not relate to the target environments that should be protected and are used by those least able to make valued judgements.

3 Groundwater Quality Standards

Contamination of groundwater has potentially serious consequences since it is widely used as a source of drinking water supply. Once contaminated it is very difficult and costly to clean up. Groundwater also provides the baseflow for streams and rivers such that it may act as the transport medium for pollutants into the surface water environment. There is a growing awareness of the role of contaminated land influencing surface water quality in our major industrialized urban areas.

The EC Groundwater Directive (80/68/EEC) is the key piece of legislation relating to the control of polluting discharges into groundwater. A group of substances (List I) must be prevented from entering groundwater, and a second group (List II) must be restricted. Exceptions can be made for discharges of substances in either list providing they are in such small quantities or concentrations as to obviate any deterioration in groundwater quality. The Directive is enforced in the UK via various legislation administered by different 'competent authorities'. With regard to contaminated land redevelopment the Local Planning Authorities are the competent authority and should control any continued leakage from developed sites by means of planning permission. Frequently, however, this duty is not recognized since many planning permissions are issued with only a perfunctory consideration of clean-up, many requirements being subsequent to the granting of permission. Where contaminated land/waste is reburied on-site then the Waste Regulatory Authority may issue a waste disposal site licence.

For operational sites there are few controls other than those that relate to specific effluent discharges (NRA consent needed). No regulations exist to control the storage or handling of hazardous chemicals in industrial premises, although Her Majesty's Inspectorate of Pollution issue authorizations for prescribed processes where BATNEEC (Best Available Techniques Not Entailing Excessive Cost) and BPEO (Best Practicable Environmental Option) must be employed. It is apparent that little attention is paid to avoiding groundwater pollution in the granting of such authorizations. There are no controls over derelict contaminated land sites, although the NRA does have powers to take action against existing or perceived pollution (Water Resources Act 1991; S.161). These are rarely used due to the lack of available resources and the uncertainty of cost recovery.

The EC Directive attaches no definitive values to the List I/II compounds. A national advisory group has been set up at the behest of the DoE under the chairmanship of the NRA. This has an objective of classifying dangerous substances that may be discharged into groundwater into either of the lists, since difficulties of interpretation have arisen with one of the categories being generic (substances that possess carcinogenic, mutagenic, or teratogenic properties). It will also consider *de minimis* concentrations.

Since groundwater is widely used as a source of drinking water and receives very little treatment before distribution to the consumer, it is often relevant to refer to drinking water standards in the absence of any other values. These are set out in the Drinking Water Regulations which enforce the EC Directive on the Quality of Water for Human Consumption.[10] However, although for some parameters maximum allowable concentrations are very low (*e.g.* individual pesticides $<0.1~\mu g\,l^{-1}$), others such as chloride (400 mg l^{-1}) and sulfate (250 mg l^{-1}) are much higher than would normally be expected in unpolluted British aquifers. Guide levels for drinking water published by the World Health Organization may therefore be a more appropriate reference in some circumstances.

4   Surface Water Quality Standards

Surface waters are not normally directly influenced by contaminated land but it is relevant to consider their quality standards since they are often the ultimate receptors of contaminants. The main EC Directive relating to surface water quality (76/464/EEC) sets out a framework for controlling pollution of surface waters by dangerous substances. As for the Groundwater Directive two lists of compounds/compound groups are defined (black list and grey list). The Directive does not in itself control the discharges since this is carried out by a series of 'daughter' Directives. The UK has approached the implementation of the European legislation by adopting Environmental Quality Standards (EQSs) for blacklist substances and Environmental Quality Objectives in relation to grey list substances for stretches of watercourse. The NRA has advised the DoE on proposals for Statutory Water Quality Objectives (SWQOs) for all controlled waters under the provisions of the Water Act 1989 (now the Water Resources Act 1991).[11] This allows for the 'use' of the particular water body to be considered. Thus the EQS for a particular substance may vary according to the SWQO of the receiving water body.

5   Adoption of Approach

The NRA has specific duties imposed upon it under the Water Resources Act 1991 to maintain and protect the quality of controlled waters. The Groundwater Directive requires all groundwater to be protected regardless of use unless the groundwater is permanently unsuitable for other uses [Article 4(2)]. This means that groundwater cannot be allowed to deteriorate from its present state. It must be demonstrated that for any potential discharge of polluting substances natural attenuation processes or physical containment will be sufficient to avoid a significant change in quality.

There are essentially two areas where guidance is needed. The first relates to the standard to which contaminated groundwater should be cleaned up. The second concerns the remediation of contaminated land where clean-up or engineering measures should be concerned with the residual impact on water resources caused by continued leaching from soils/waste remaining on site. These are considered further below.

6   Contaminated Groundwater

Historical Diffuse Pollution

The NRA's Groundwater Protection Policy[3] states, in Policy D6, "In areas where historical industrial development is known to have caused widespread groundwater contamination, the NRA will review the merits and feasibility of groundwater clean-up depending upon local circumstances and available funding". This statement is designed to cover situations where many point-source discharges, over a long period, have combined to cause a diffuse pollution effect whereby no one polluter could be identified with certainty and without excessive cost. Examples can be found in many major industrialized urban areas of the UK.[12] It may also be that a particular polluting industry has ceased to operate and that the present site occupier uses different processes and materials. In such instances we cannot aim to return an aquifer to a pre-industrial state. Instead, where water is required from such urban areas, treatment at the point of abstraction must be adopted. There are 14 public supply abstractions in the UK currently affected by chlorinated solvents from industrial sources sufficiently to warrant treatment.[13] However, the contamination will have been caused in most cases by activities that were not subject to pollution control legislation and before general awareness of safe practices to avoid groundwater pollution.

Point Source Pollution/Potable Abstraction

Where the pollution is discrete and the polluter/polluting event known, then the targets must be identified. These targets may be individual abstractions, springs, or watercourses into which the groundwater is discharging. An assessment should then be made of the degree of risk to each target and the extent and timescale of aquifer rehabilitation tailored to fit the specific case. For example, a public supply abstraction directly down-gradient of the pollution source would be considered a high risk, particularly if travel times are fast. Provision would need to be made for interception of the contaminated plume followed by treatment and re-injection or discharge to sewer/watercourse depending upon circumstance.

Such assessments often involve the use of modelling techniques to estimate travel times and consequential concentrations at the identified targets. However, models are only tools that are to be used with caution, especially where data are limited, which is invariably the case. Where there is a degree of confidence in the model it can be highly influential in determining remedial action.

Point Source Pollution/Limited Resource Potential

In situations where no targets are identified and groundwater is considered to be of little value as a resource then the extent to which clean-up is necessary may be limited. Such a situation may arise where a shallow pocket of gravels rests on clays. The groundwater within the gravels will be restricted in extent and hence has no resource potential. It may also have no immediate outlet to surface water. In this case it may be acceptable to leave the groundwater to natural degradation processes, although it must be recognized that these will only work over long periods of time. Moreover, it will be necessary both to demonstrate the low level of environmental risk and to show that natural attenuation processes will occur before such an approach can be adopted.

Discussion on Contaminated Groundwater

The last two situations are at opposite ends of a spectrum. In the first instance the ultimate clean-up standard to be achieved is that of drinking water quality, particularly where treated water is re-injected into the aquifer. Where it can be proved that dilution/attenuation will take account of residual trace contaminants a more flexible approach could be adopted. In practice this will inevitably have to be accepted for most organic pollution instances since current technology is unable to clean-up groundwater sufficiently without excessive costs. Indeed an approach based on a combination of clean-up and treatment at the point of abstraction may have to be accepted as the most cost-effective solution.

In the second example there is no particular standard set, although the objective is a long-term improvement in groundwater quality. Intermediate cases where the abstraction at risk is not used for potable supply, or where the groundwater is as yet underutilized, have to be considered on a case by case basis. Since drinking water is the primary 'use' for groundwater these standards must always be borne in mind, although it may not be necessary, practicable, or economically possible to achieve this in every case. For instance, natural groundwater quality in some geological strata is poor and would not meet the standards laid down in the Drinking Water Regulations[10] for certain parameters. In such circumstances it would not be sensible to require remediation standards better than those that pre-existed the pollution event. Often the groundwater quality will be largely unknown since comprehensive monitoring networks have yet to be established for all aquifers. This presents some difficulty therefore in setting baseline standards. Situations where groundwater ultimately discharges to surface waters are discussed further in the next section.

Of course all cases imply that the original source of contamination has been removed or addressed so that continuing pollution will not occur. Continuing pollution would be illegal under the terms of the Groundwater Directive.

7   Clean-up Standards for Soils

Contaminants in soils may be leached into groundwater if continually exposed to precipitation or other means of mobilization (soakaways, drains, *etc.*). The issue for groundwater protection does not relate so much to the degree to which soils should be cleaned up but to what level leaching can be reduced. The flux of soluble contaminants to groundwater from soils can be reduced by excavation and removal off-site, by chemical/physical treatment, or by various engineering methodologies, for example simple capping of the site.

In each case we need to be able to assess the risks to the particular water body in need of protection and set maximum levels for leached contaminants accordingly. This can be carried out using impact assessment techniques that quantify the hazard, the pathway, and the potential impact on the defined target using data obtained from the site investigation. Impact assessments therefore provide a more rational way forward than generic sets of standards.

If the underlying groundwater is a potential resource for drinking water supply, then it becomes the target and the level of leached contaminants must relate to the necessity for drinking water standards not to be exceeded in that resource. UK law has adopted a wider definition of groundwater than that in the Groundwater Directive. It includes "any water contained in underground strata",[14] so the legislation could be interpreted as requiring that all water draining from the site should be of drinking water quality. However, for all practical purposes protection is given to the water in the saturated zone (*i.e.* the European definition) and therefore due account can be taken of attenuation processes, where quantifiable, in the unsaturated zone (the pathway). This issue is the subject of a current review by the NRA/DoE committee set up to determine the status of specific substances with respect to Lists I/II of the Groundwater Directive.

Where groundwater is not the primary concern because of poor natural water quality, historic pollution, or low resource potential, then the primary target water body may be the surface water into which the groundwater discharges. In this case groundwater becomes the pathway. An example of this approach is outlined below.

Example

A contaminated site of a former steel works is situated on alluvium (sands and silty clays) overlying Coal Measures in an historically industrialized urban area. There are no groundwater abstractions in the area and groundwater resources are not considered significant. A tributary stream, draining to the headwaters of the main watercourse for the area, runs adjacent to the site and groundwater in the alluvium is considered to be in hydraulic continuity with this stream. Groundwater quality data is available for a few locations in the area but is variable. It reflects both industrial activity and naturally high mineralization. Surface water quality data is also variable reflecting ground water quality to some extent, especially in dry weather. There have been considerable efforts to improve the quality of the main river in addressing direct discharges from both industry and sewage works. These have met with some success.

Although no specific water quality objectives have been set for the watercourses, new and revised consent standards are worked out using EQSs. The EQSs relate to the use of the main river as an improving coarse fishery and to achieving a higher general recreational profile as former industrial land along the riverbank is regenerated and made more accessible to the public. Little is understood about the geochemistry of the alluvium or the general hydrogeology of the site and so no assessment can be made of attenuation mechanisms. Hence the quality of leachate that can be allowed to continue to drain from the site after remediation should be related to the EQSs.

Leachability Tests

Adopting this approach presupposes that data will be available for leachate quality. Often it is not, since many site investigations only include analyses for total concentrations within the samples taken. It is therefore a prerequisite of this approach that leachability tests are carried out on a representative number of samples. There are many leachability tests available of which most have been used to assess materials for landfilling purposes. Many are undertaken under aggressive conditions to provide for 'worst case' scenarios in landfill environments and may therefore be inappropriate in relation to contaminated land situations. Considerable difficulty has also been experienced in interpretation of data resulting from a wide range of tests and so the NRA has commissioned research to produce a standard methodology. The work is presently under review with regard to both validation and compatibility with methods being considered by the European CEN Technical Committee working on leaching tests for waste acceptance and characterization.

The NRA method is intended to be a basic test, recognizing the difficulties of obtaining representative samples on contaminated land sites. It is not important to have an extremely accurate test with high reproducibility when there are many other uncertainties involved in adopting an impact-based assessment approach.

Data Interpretation

Once leachate analyses have been undertaken, the data can be compared with the EQSs of the watercourse being protected. Where potential leachate concentrations exceed the EQSs, consideration should be given to various options to limit the residual impact of the redeveloped land. These may be to: (a) investigate the potential for attenuation between the site and watercourse; (b) engineer the site so that infiltration is further reduced; (c) carry out further treatment of the contaminated materials so that the concentration of leachable contaminants is reduced; or (d) construct barriers to contain the contamination. Any, or combinations, of these options may have a part to play.

Such assessment of the problems at any sites must be carried out well ahead of planning applications or other authorizations so that the regulatory authorities can consider the impact of proposals properly. All too often the issues relating to pollution are left to the development stage when budgets are set and options limited. Careful early planning may lead to less delay in the long term and less residual liability, and may even avoid abortive expenditure.

The impact assessment approach can be adopted for specific sites or on a catchment-wide basis. Where there are many sources of pollution in the locality it is only reasonable to consider a catchment approach. Therefore, the same set of EQSs can be used for all sites within the catchment of a particular watercourse. This provides a degree of consistency within a given area although should a less stringent value for leached contaminants be required site-specific data will always be necessary to show that the risks are reduced by attenuation or other means. This approach is currently being evaluated for that part of the Black Country area of the West Midlands that lies within the catchment of the headwaters of the River Tame.

8 Conclusions

Existing guidance on clean-up standards for the water environment is lacking in the UK. Numeric standards have been adopted on an *ad hoc* basis that were never intended for the uses to which they are being put.

Standards do not exist for groundwater since its quality is variable and, because of inadequate monitoring, unknown for many areas. Since groundwater is widely used as a source of potable supply, drinking water standards are often quoted as objectives for clean-up.

Where groundwater resources in major aquifer areas are contaminated, drinking water standards are the most appropriate to use. However, it may be appropriate to set higher values where natural attenuation is anticipated, or lower values in those circumstances where maximum drinking water standards are considerably in excess of background values.

Surface water quality standards are a more appropriate measure for clean-up where groundwater is of less importance as a resource, but is a significant component of stream baseflows. This is a situation that frequently occurs in many industrialized areas of the UK.

The adoption of an impact assessment approach to contaminated land remediation will provide a better method of deriving clean-up objectives for the purposes of protecting the water environment than the use of generic standards. However, standards do have merit where used for specific catchments and relate to surface water quality. The use of standard leachability testing methodology and data application will be an aid to adopting this approach in the future. It is intended to adopt these within the NRA.

References

1 R. J. Flavin and R. C. Harris, 'Contaminated land: implications for water', *J. Inst. Water Environ. Manage.*, 1991, **5**, 529–533.
2 R. C. Harris and G. Gates, 'Developments in NRA policy and practice', *Environ. Policy Practice*, 1993, **3**, 113–124.
3 National Rivers Authority (NRA), 'Groundwater Protection Policy, Bristol, 1992.
4 Interdepartmental Committee for the Redevelopment of Contaminated Land, 'Acceptable Levels of Toxic Elements in Soils', ICRCL, DoE, London, 1979.
5 M. A. Smith, 'Experiences of the development and application of guidelines for contaminated sites in the United Kingdom', in 'Proceedings of Developing Clean-Up Standards for Contaminated Soils, Sediment and Groundwater', January 10–13, 1993, Water Environment Federation, USA, 1994, pp. 195–204.
6 Interdepartmental Committee for the Redevelopment of Contaminated Land, 'Guidance on the Assessment and Redevelopment of Contaminated Land', ICRCL Guidance Note 59/83, 2nd edn, DoE, London, 1987.
7 R. T. Kelly, 'Site investigation and materials Problems', in 'Proc. Conference on Reclamation of Contaminated Land', Eastbourne, 1979, B2/1-14, Society of Chemical Industry, London, 1980.
8 Ministry of Housing Physical Planning and the Environment, 'Guidelines for Soil Sanitation', The Netherlands, 1987.

9 J.J. Vegter, 'Development of soil and groundwater clean-up standards in the Netherlands', in 'Proceedings of Developing Clean-Up Standards for Contaminated Soils, Sediment and Groundwater', January 10–13, 1993, Water Environment Federation, USA, 1994, pp. 81–92.

10 Department of the Environment, 'The Water Supply (Water Quality) Regulations', Statutory Instrument 1989 No. 1147, HMSO, London, 1989.

11 National Rivers Authority (NRA), 'Proposals for Statutory Water Quality Objectives', Water Quality Series No. 5, Bristol, 1992.

12 D. N. Lerner and J. H. Tellam, 'The protection of urban groundwater from pollution', *J. Inst. Water Environ. Manage.*, 1992, **6**, 28–37.

13 R.C. Harris, 'Groundwater pollution risks from underground storage tanks', *Land Contam. Reclamation*, 1993, **1**, 197–200.

14 Water Resources Act 1991, HMSO, London, 1991.

Chapter 10

OPRA (Operator and Pollution Risk Appraisal): A Practical System for Rating and Managing Environmental Risks from Industrial Processes

By Huw Jones

DNV INDUSTRY, PALACE HOUSE, 3 CATHEDRAL STREET, LONDON SE1 9DE, UK

1 Background

Once an industrial process is authorized under Integrated Pollution Control (IPC), Her Majesty's Inspectorate of Pollution (HMIP) inspectors must ensure that the conditions of authorizations are continuously complied with by means of inspection and monitoring. The Operator and Pollutions Risk Appraisal (OPRA) was developed to help HMIP target its resources to maximum effect in enforcing compliance with regulations and protecting the environment in the most cost-effective manner. The Environmental Agency has inherited OPRA in taking over HMIP's regulatory role in April 1996.

The OPRA system was developed within the Operations Division of HMIP with strong support from the HMIP Advisory Committee, which advocated the use of objective and transparent 'performance indicators' to assess industry and regulatory effectiveness. The committee believed that the system was a valuable tool for making the thinking behind visit frequencies transparent, that it would send a message to operators on how well they were performing, and that it needed to be a joint scheme to work effectively.

The testing of OPRA was implemented by all regions of HMIP within a very limited timescale, and statistical analysis was performed on the trial data gathered from inspectors. An independent review of OPRA, carried out by Professor A. F. M. Smith of Imperial College, concluded that OPRA "...can indeed provide the basis of a scientifically defensible procedure".

Publication of a consultation document on the OPRA system ('Operator and Pollution Risk Appraisal', HMIP, April 1995) was agreed by the Secretary of State for Environment, and completed, along with a press release, at the beginning of April. Since this time, there has been considerable and widespread interest in the scheme. Over 3500 copies of the consultation document have been distributed to industry, regulators, and specialist organizations.

2 System Description

The OPRA system provides a mechanism for rating an individual process by assessing the pollution hazard of the process (Pollution Hazard Appraisal, PHA) and the quality of the operator's performance in managing that process (Operator Performance Appraisal, OPA). The balance between OPA and PHA scores is used as an indicator of the environmental risk of the process. This information may then be used for a variety of purposes, such as setting inspection frequencies and identifying potential areas for improvement. Figure 1 shows the overall approach of OPRA.

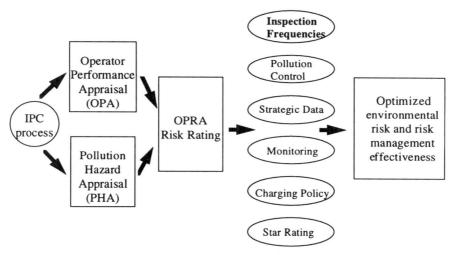

Figure 1   *Operator and pollution risk appraisal*

The primary objective of OPRA is to optimize regulatory/operator risk management effectiveness. This will be achieved by ensuring that OPRA identifies the major sources of risk and thus allows HMIP's and the operator's risk management efforts to be focused and effective. This concept is illustrated in the context of inspection in Figure 2, which shows that using a risk-based approach should result in a lower risk for the same level of inspection activity, or equally should enable a reduction in inspection activity without compromising risk levels.

The two main assessment procedures within OPRA, OPA, and PHA involve a straightforward evaluation and rating of 14 key indicators of the operator performance and pollution hazard of the process. The OPA indicators are as follows:

- recording and use of information
- knowledge and implementation of authorization requirements
- plant maintenance
- management and training

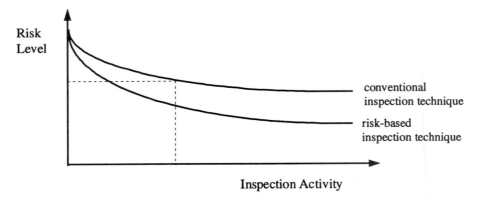

Figure 2   *Risk-based inspection*

- plant operation
- incidents, complaints, and non-compliance
- auditable environmental management systems

Each of these attributes are evaluated for the process as a whole and assigned a score between 1 and 5. The scores are then multiplied by a weighting factor to reflect the importance of each attribute, and the scores added together to obtain the overall OPA score. The PHA score is obtained in a similar fashion, by evaluation and rating of the following attributes:

- presence of hazardous substances
- scale of use of hazardous substances
- frequency and nature of hazardous operations
- techniques for hazard prevention and minimization
- techniques for hazard abatement
- location of process
- offensive substances

3 Further Development and Implementation of OPRA

OPRA has been the subject of extensive consultation within and outside of HMIP, and responses to the consultation document indicate that there is a high level of support for the OPRA concept and its objectives. There is a general desire to ensure that the dialogue between industry and HMIP is as transparent and constructive as possible, and that risk management resources are allocated cost-effectively. Industry would like to see a system that reflects the 'polluter pays' principle and rewards effort to improve environmental performance.

The consultation process has raised a number of issues, however, that HMIP intends to address in the further development of OPRA. The main concerns are:

(1) The need for consistent evaluation of relative pollution risks from different industry sectors and sites.

(2) There is substantial opposition to the publication of OPRA results, on the grounds that the approach is not yet proved and the results may mislead the public.

(3) Linking OPRA to important issues such as charging should be approached with care and only be implemented once the system is established.

(4) Additional guidance may be required, while ensuring that the appropriate degree of simplicity and professional judgement is maintained.

HMIP is further developing OPRA in the light of these views, with the intention of formally implementing the system in early 1996. As indicated in Figure 1, there are a number of applications for which OPRA results could be used:

- Inspection frequencies: planning and allocation of inspector resources to target the more important sources of environmental risk and relax inspection effort where there is a low risk due to the inherent low hazard or high performance of the operator.
- Pollution control: in support of dialogue between HMIP and operators, to enhance the level of pollution control where weaknesses are identified, and thus establish a higher overall standard of environmental performance across industry.
- Strategic data: to provide HMIP with data on overall performance of different industry sectors and geographical regions, and to establish trends in environmental risk and performance over time.
- Monitoring: to assist HMIP and the operator to determine the appropriate type and level of monitoring of releases required to provide the necessary information to optimize environmental performance.
- Charging policy: linking of HMIP charges to operators via the OPRA score, to reflect more accurately the level of performance by the operator and the effort required of HMIP to ensure compliance with the objectives of IPC.
- Star rating: publication of the overall results of OPRA scores for industry, to provide an incentive for low performers to approach the levels of performance set by the best operators.

HMIP's near-term development strategy is likely to focus on applying OPRA to setting inspection frequencies and discussing pollution control issues with operators. Extension of the OPRA system to other areas may be considered once the system is fully established within HMIP.

Chapter 11

Site-Specific Considerations in Risk Assessment

By Holly A. Hattemer-Frey and Virginia Lau

DAMES & MOORE, 633 17TH STREET, SUITE 2500,
DENVER CO 80202-3625, USA

1 Introduction

Risk assessment is a process that synthesizes available data on exposure and toxicity of chemicals and uses scientific judgement to estimate the potential risk to human health and the environment. Uncertainties in the risk assessment process often arise when risk assessors are forced to make assumptions about the nature and extent of exposures in the absence of site-specific data. The US Environmental Protection Agency (EPA) encourages the collection of site-specific data so that risks can be accurately assessed on a case-by-case basis.[1,2] Site-specific data should be used to the maximal extent possible to:

- reduce or eliminate reliance on standard default values or other default data taken from the literature; and
- reduce uncertainty associated with estimating exposures and risks.

Collecting site-specific data will often lower exposure and risk estimates, which will, in turn, result in lower remediation costs, since it avoids the need to make conservative assumptions. This chapter describes the importance of using site-specific data to prepare Human Health Risk Assessments (HHRAs) and illustrates the benefits of using site-specific versus default data through various case studies.

HHRAs characterize the nature and extent of potential adverse human health effects associated with exposure to site-related contamination. One of the first steps in the HHRA process is to determine the location and number of media samples needed to support HHRA calculations. Once site-specific data have been collected and analysed, chemicals of potential concern (COPCs) are identified. This step is followed by an evaluation of potential exposure pathways and the quantification of chronic daily intakes. The identification of exposure routes (*e.g.* inhalation, ingestion, and dermal contact) and receptor locations is crucial to determine the validity of potential exposure pathways. After complete exposure pathways have been identified, exposure-point concentrations and receptor intakes are calculated. The next step, toxicity assessment, identifies COPCs that may cause adverse health effects in exposed populations. Risk characterization combines results of the exposure and toxicity assessments to yield quantitative estimates of risk. Finally, a qualitative or quantitative analysis of uncertainty, which summarizes potential factors that may have resulted in under- or overestimates of risk, is a final component of the risk assessment process.

HHRAs for contaminated sites span a continuum of complexity and detail. At each site, there is a highly variable ability to obtain reasonable and reliable site-specific data for various human activity and behaviour patterns. It is frequently necessary to use model-predicted exposure-point concentrations because data either do not exist or they are difficult to obtain in a systematic, cost-effective fashion. For example, since it can be costly and difficult to collect site-specific ambient air data, ground-level air concentrations are often modelled. Despite these limitations, a substantial amount of site-specific data can be collected, which can greatly reduce uncertainty in exposure and risk estimates.

2 Physico-chemical Properties

Many parameters, such as water solubility, vapour pressure, octanol–water partitioning, and bioaccumulation, influence the behaviour and fate of chemicals released into the environment. The importance of these factors and how they influence each other is often inadequately understood. However, the examination of a few basic physico-chemical properties can provide insight into the behaviour and fate of chemicals released into the environment and the potential long-term persistence of the chemical in a specific medium.

Water Solubility

Water solubility is the maximum amount of a chemical that will dissolve in pure water at a specific temperature and pH. A chemical's solubility in water affects its fate and transport in all environmental media, and significantly influences human exposure through aquatic pathways. Highly soluble compounds tend to leach rapidly from soil into groundwater and surface water supplies. In addition, they tend to be less volatile,[3] more biodegradable,[4] and more mobile[5] than less soluble chemicals. The solubility of organic chemicals can vary dramatically depending on pH, redox potential, and the types and concentrations of organic complexing species found in the soil. Solubilities can range from less than 1 ppb to more than 100 000 ppm, with most common organic constituents having a solubility between 1 and 100 000 ppm.

Vapour Pressure

Vapour pressure measures the relative volatility of a chemical in its pure state, which is useful for determining the extent to which a chemical will be transported into air from soil and water surfaces. Volatilization is a major route for the distribution of many chemicals in the environment.[6] A chemical's volatility is affected by its solubility, vapour pressure, and molecular weight, as well as the nature of the air-to-water or soil-to-water interface through which the chemical must pass.[4] For example, chemicals that have a low vapour pressure and a high affinity for soil or water are less likely to vaporize than chemicals that have a high vapour pressure and a weak affinity for soil or water.

Henry's Law Constant

Henry's law constant (H) combines vapour pressure with solubility and molecular weight and is useful for estimating releases to air from water. It is defined as the ratio of the partial pressure of a compound in the vapour phase to the concentration in the liquid phase (mole fraction), or the ratio of a chemical's vapour pressure to its solubility. Both vapour pressure and solubility are temperature dependent. Under equilibrium conditions, Henry's law describes the proportion of the chemical in the gas phase above the liquid to the concentration of the chemical in the liquid. A dimensionless form of this parameter may be expressed by dividing Henry's constant (atm m^3 mol^{-1}) by the universal gas constant (8.25×10^{-5} atm m^3 mol^{-1} K^{-1}) and the ambient temperature (293 K). Henry's law constant is significantly affected by the temperature and chemical composition of the water.[7,8] For instance, a 10 K increase in temperature will result in an approximately three-fold increase in Henry's law constant for volatile hydrocarbons. Dissolution of chemicals in seawater will generally enhance the vapour pressure such that the Henry's law constant will increase to some degree.[8]

Bioconcentration/Biotransfer Factors

Assessing the environmental fate of chemicals depends largely on being able to predict the extent to which they will bioaccumulate in living organisms, including fish, cattle, and plants. Organisms can concentrate chemicals in their tissues at levels substantially higher than a chemical's concentration in water (in the case of aquatic organisms), in food (in the case of terrestrial organisms), or in soil (for terrestrial plants). A traditional measure of a chemical's potential to accumulate in biota is the bioconcentration factor (BCF), which is defined as the steady-state concentration of a chemical in an organism or tissue (mg kg^{-1}) divided by the steady-state concentration of it in water (for aquatic organisms), soil (for vegetation), or food (for terrestrial organisms). This concept, however, is not readily applied to humans, because the amount of chemical in various food items of the human diet can vary markedly. Hence, for risk assessment purposes, it is more efficacious to examine the biotransfer factor (BTF), which is defined as the steady-state concentration of organic material in an organism or tissue (mg kg^{-1}) divided by the average daily intake of organic material (mg day^{-1}).[9]

Octanol–Water Partition Coefficient (K_{OW})

Organisms tend to accumulate chemicals in the lipid portions of their tissues. Thus, one way to determine the bioaccumulation potential of a chemical is to measure how lipophilic it is. Since it is difficult to directly measure a chemical's lipophilicity, researchers typically use the octanol–water partition coefficient (K_{OW}) to predict a chemical's tendency to partition between an octanol component (a good surrogate for fat) and water. The octanol–water partition coefficient is defined as the ratio of a chemical's concentration in the octanol phase to its concentration in the aqueous phase.[4] It is directly related to a chemical's tendency to bioconcentrate in biota[10–16] and inversely correlated with water solubility.[10,15] Hence, the octanol–water partition coefficient is used extensively to estimate the bioconcentration potential of organics in biological systems. Chemicals with large K_{OW} values tend to accumulate in soil, sediment, and biota but not water. For example, lipophilic compounds such as dioxin, 1,1,1-trichloro-2,2-bis(*p*-chlorophenyl)ethane (DDT), and polychlorinated biphenyls, are most soluble in organic matter. This class of chemicals tend to bioaccumulate in biota and vegetation, sorb strongly onto soil, sediment, and vegetation, and transfer to humans through the food chain. Conversely, chemicals with low K_{OW} values tend to partition mostly into air or water. For example, volatile organics such as trichloroethylene and tetrachloroethylene tend to be widely distributed in air with inhalation being the primary pathway of human exposure.

Partition Coefficients

Partition coefficients are empirical constants used to describe the distribution of a chemical between two media. For instance, the migration of chemicals through the soil may be slowed by the sorption of contaminant to soil particulates. Many compounds form a hydrophobic bond with organic matter and mineral groups present on the surface of soil particles. The organic carbon partition coefficient (K_{OC}) is a measure of the relative sorption potential for organic chemicals. K_{OC} depicts the chemical's tendency to adsorb to soil and is independent of soil properties. K_{OC} is the ratio of the amount of chemical adsorbed per unit weight of organic carbon to the chemical concentration in solution at equilibrium as indicated in Equation (1):

$$K_{OC} = \frac{\text{mg adsorbed } / \text{ kg of organic carbon}}{\text{mg dissolved } / \text{ l of solution}} \quad (1)$$

K_{OC} values can range from 1 to $10^7 \, \text{l kg}^{-1}$, with larger values indicating a greater potential for adsorption. K_{OC} is directly indicative of soil and sediment sorption, which are often significant chemical fate processes at contaminated sites.

The significance of K_{OC} values varies with different potential exposure pathways. For groundwater, chemicals with low K_{OC} values tend to leach faster from a waste source to an aquifer and have relatively rapid movement through the water (*i.e.* limited retardation of the chemical). Conversely, a high K_{OC} value indicates strong binding to soil or sediment, which means that less of the chemical will be dissolved in run-off but also implies that run-off of contaminated soils may occur over a long time frame.

The extent of sorption of most organic compounds may be reasonably estimated from the organic carbon content of the soil. The soil water partition coefficient (K_d) is a site-specific measure of the tendency of a chemical to be adsorbed by soil or sediment. Since almost all of the adsorption of organic chemicals by soil is governed by the organic carbon content in the soil, the organic carbon partition coefficient and the fraction of organic carbon (f_{OC}) may be used to estimate the soil–water partition coefficient. The relationship is expressed as $K_d = K_{OC} \times f_{OC}$.

Although it is fairly easy to calculate K_d values for organics, K_d estimates for metals are highly variable and susceptible to environmental changes. For instance, laboratory measurements of K_d values using either a column or batch method will generally introduce biases into the value reported since neither method accounts for non-equilibrium conditions or the influences of convection and diffusion. Since K_d is a parameter defined by the ratio of two concentrations, a small error in the measurement of either the soil or water concentration may produce significant errors in the ratio. In addition, K_d estimates for metals are highly sensitive to soil properties including pH, clay content, organic matter content, free iron and manganous oxide contents, and particle size distribution.[17] Because of the inherent uncertainties associated with producing K_d values for metals, K_d estimates are normally presented in the literature as ranges that may encompass several orders of magnitude. Table 1 presents K_d values for a selected number of metals normally encountered in the environment.

Diffusion Coefficients

Diffusion is the process by which a chemical moves from areas of high concentration to areas of lower concentration. The amount of a contaminant passing through a given area per unit time is controlled by Fick's Law:

$$J = -D \left(\frac{\mathrm{d}C}{\mathrm{d}x} \right) \qquad (2)$$

where J = chemical flux (mg cm^{-2} s^{-1}); D = chemical-specific diffusion coefficient (cm^2 s^{-1}); C = concentration (mg cm^{-3}); and x = length in the direction of movement (cm). Diffusion rates for chemicals may be estimated for both air and aqueous solutions. The rate at which a chemical diffuses through a medium is directly related to the cross-sectional size of the molecule and its mean free path (*i.e.* the distance travelled without hitting another molecule). Diffusivities of molecules in air are usually of the order of 0.1 cm^2 s^{-1}, while diffusivities in water are approximately 5×10^{-6} cm^2 s^{-1}.

Table 1   *Range of measured* K_d *values for selected metals*

Metal	Observed range of soil–water partition $(K_d)^a$ (ml g^{-1})
Arsenic	1.0–8.3
Cadmium	1.26–26.8
Chromium(III)	470–150000
Copper	1.4–333
Iron	1.4–1000
Lead	4.5–7640
Magnesium	1.6–13.5
Potassium	2–9
Selenium	1.2–8.6
Zinc	0.1–8000

^{a} Reference 17.

Variability in Fate and Transfer Coefficients

Table 2 lists log K_{OW}, K_{OC}, Henry's law constant, vapour pressure, solubility, air diffusion, and fish BCFs for 13 organic chemicals. Table 2 shows that for some chemicals, there is tremendous variability for some of these parameters. For example, Henry's law constants, K_{OW} values, K_{OC} values, solubility, and vapour pressures reported in the literature for di-*n*-butyl phthalate vary by a factor of 16, 11, 10, 350, and 6, respectively. The variability of fate and transport coefficients is the result of temperature and pressure differences that can occur under laboratory conditions when the value is measured. Almost all chemical factors are sensitive to environmental conditions such that site-specific modification of these factors could occasionally be warranted. For an initial screening risk assessment, however, the factors measured under ambient conditions should be used.

3   Data Collection and Evaluation

Sampling data are needed to characterize the nature and extent of contamination present in environmental media. The needs of the HHRA must be considered early in the development of sampling plans to ensure that samples collected are suitable for its purposes. The site-specific data needed to support a quantitative HHRA include:

- chemical concentrations in key sources and media of interest,
- chemical concentrations in site-specific baseline samples for all relevant media,
- source characterization, especially information related to chemical release potential,
- samples defining the vertical and aerial extent of contamination, and
- samples needed to determine the fate, transport, and persistence of site-related chemicals in environmental media.

Table 2 Range of physicochemical properties for 13 select organic chemicals

Chemical	Log K_{OC}	K_{OC} (ml g^{-1})	Henry's Law Constant (dimensionless)	Diffusion coefficient for air (cm^2 s^{-1})	Solubility (mg l^{-1})	Vapour pressure (mmHg)	Bioconcentration factor for fish
Benzene	2.13 [a,h] 2.12 [b,d] 2.1 [l,m]	65 [a] 83 [b,d] 60 [l]	5.40×10^{-3} [a] 5.59×10^{-3} [d] 6.6×10^{-3} [c]	0.0923 [b,i,a,b] 0.088 [c] 0.0932 [j]	1780 [a,b] 1750 [d]	76 [a] 95.2 [b,d] 96.2 [c]	6.5 [a] 5.2 [b,d]
Carbon tetrachloride	2.83 [a] 2.64 [b,d,h,m]	110 [b,d] 439 [a]	2.00×10^{-2} [a] 2.41×10^{-2} [d] 3.0×10^{-2} [c]	0.0845 [b,i] 0.078 [c] 0.0828 [j]	800 [a,b] 757 [d]	91.3 [a] 90 [b,d] 113 [c]	30 [a] 19 [b,d]
Chlorobenzene	2.84 [a,b] 3.79 [m]	333 [a,b]	3.46×10^{-3} [a] 3.93×10^{-3} [c]	0.07627 [b] 0.073 [c]	490 [a] 500 [b]	8.8 [a] 11.7 [b] 11.8 [c]	33 [a] 10 [b]
1,1-Dichloroethane	1.79 [b,d] 2.13 [a]	30 [a,b,d]	5.7×10^{-3} [a] 4.31×10^{-3} [d]	0.0964 [b] 0.0826 [j]	5500 [a,b,d]	182 [a,b,d] 591 [c]	2.9 [a]
1,1-Dichloroethylene	2.13 [a] 1.84 [d]	65 [a,b,d]	0.154 [a] 0.0154 [c] 3.4×10^{-2} [d]	0.08386 [b]	400 [a] 2250 [b,d]	500 [a] 600 [b,d] 630 [c]	6.4 [a] 5.6 [b,d]
Di-n-butyl phthalate	4.57 [a] 5.6 [d] 5.3 [e] 5.2 [g]	17900 [a] 170000 [d]	4.5×10^{-6} [a] 2.82×10^{-7} [c,d]	0.0438 [c] 0.0421 [j]	4500 [a] 13 [d]	1.6×10^{-4} [a] 1.0×10^{-3} [d] 1.0×10^{-4} [c]	1800 [a]

Ethylbenzene	3.15 [a,b,d]	1100 [b,d] 681 [a]	6.43×10^{-3} [c,d] 7.9×10^{-3} [a]	0.0707 [b,i] 0.075 [c] 0.0755 [j]	152 [a,b,d]	7 [a,b,d] 10 [c]	95 [a] 37.7 [b,d]
Naphthalene	3.3 [a] 3.36 [l] 3.59 [m]	962 [a] 870 [l]	4.82×10^{-4} [a,c]	0.069 [c]	31.7 [a]	0.053 [a] 0.23 [b]	44–77 [a]
2,3,7,8-TCDD (Dioxin)	6.72 [b,d] 6.84 [f] 6.42 [k]	2.63×10^{7} [k] 3.3×10^{6} [b,d]	3.6×10^{-3} 1.6×10^{-5} [k]		2×10^{-4} [b,d] 7.91×10^{-6} [k]	1.7×10^{-6} [d] 7.4×10^{-10} [c] 7.4×10^{-10} [k]	5000 [b,d]
Tetrachloroethylene	3.14 [a] 2.6 [d,h] 2.53 [l]	665 [a] 364 [b,d,l]	0.0227 [a] 0.024 [c] 2.59×10^{-2} [d]	0.072 [c] 0.07852 [i] 0.07729 [b] 0.0797 [j]	150 [a,b,d] 17.8 [d]	14 [a] 31 [b,d] 19 [c] 60 [b]	49 [a]
Toluene	2.73 [a,b] 2.69 [h]	259 [a] 300 [b]	6.61×10^{-3} [a] 6.68×10^{-3} [c]	0.067 [c] 0.08301 [b,i] 0.0849 [j]	515 [a,b]	22 [a,b] 36 [c]	27 [a] 10.7 [b]
Vinyl chloride	1.23 [a] 1.38 [b,d]	8.2 [a] 57 [b,d]	0.965 [a] 8.62×10^{-2} [c] 8.19×10^{-2} [d]	0.11375 [i] 0.106 [c]	1100 [a] 2670 [d] 4270 [b]	2300 [a] 2660 [b,c,d]	0.8 [a] 1.2 [b,d]
Xylenes	3.16 [a] 3.26 [b,d]	240 [b,d] 691 [a]	6.32×10^{-3} [a] 7.04×10^{-3} [d] 6.25×10^{-3}	0.07597 [i]	0.3 [a] 198 [b,d]	7–9 [a] 10 [b,d] 8.5 [c]	70 [a]

[a] Reference 18; [b] reference 7; [c] reference 19; [d] reference 20; [e] reference 21; [f] reference 22; [g] reference 23; [h] reference 10; [i] reference 24; [j] reference 25; [k] reference 26; [l] reference 27; [m] reference 28.

These data are necessary to identify COPCs for all media, to quantify exposure-point concentrations, and to evaluate if current and future media concentrations are expected to increase or decrease substantially over time. It is also necessary to conduct a preliminary assessment of potential human exposure pathways to ensure that all site-specific data needed for the HHRA are collected during field investigations and to avoid superfluous sample collection. For example, if current and future land use at a site is expected to be commercial/industrial, it is not necessary to collect data on food items that may be grown or produced on-site.

Contaminant release, fate, and transport models are often needed to supplement monitoring data to calculate exposure-point concentrations. The preliminary identification of modelling requirements will ensure that data needed to calibrate and validate the necessary models are collected along with the specific physical and chemical data needed to run the models. Baseline or background sampling data are needed to distinguish site-related contamination from naturally occurring and anthropogenic levels. For example, many pesticides, most of which are not naturally occurring, may be ubiquitous in certain (*e.g.* agricultural) areas, and levels measured on-site may not be attributable to site releases. Baseline samples should be collected in areas that have not been influenced by site releases but have the same basic hydrogeologic characteristics. Areas directly upwind, upstream, or upgradient of the affected area often offer suitable locations for baseline sample collection.

The media of concern at a given site are generally any currently contaminated media to which individuals may be directly or indirectly exposed and any currently uncontaminated media that could become contaminated in the future. In cases where contamination has not yet reached points of human exposure but could be transported there in the future (*e.g.* future contamination of groundwater from percolation through already-contaminated soil), samples between the contaminant source and potential exposure locations should be taken to facilitate evaluation of potential future exposures.

4   Site-Specific Soil Data Needs

Contaminated soil has the potential for direct human contact and in many cases is the primary source of contaminant transport to other media. The heterogeneous nature of on-site contamination may make it difficult to ensure that representative soil samples are collected. Sample depth should be based on potential exposures and transport routes of concern. If individuals currently live on or near the site, surface soil samples should be collected so that direct ingestion of and dermal contact with COPCs in soil can be evaluated. Similarly, home-grown produce samples should be collected, since many metal and organic constituents are readily taken up through plant roots. If groundwater may be impacted by contaminated soil, deeper soil samples may need to be collected.

Fate and Transport of Organics in Soil

The fate and transport of contaminants is influenced by the physical, chemical, and biological processes that interact with the natural system. The governing mechanisms that determine the fate and transport of organic compounds in soil include volatilization by molecular diffusion, adsorption of soluble contaminants onto organic matter, leaching of contaminants by water infiltration, and natural decay and degradation. Dependent on their physicochemical properties, organics may readily volatilize by vapour-phase diffusion through air-filled pore spaces within the soil column. Some percentage of the organic compound may also attach to organic matter and remain in the soil. Rainfall and surface water infiltration may leach organics through the soil column into ground water or slough them downgradient. Depending on the half-life of the compound, natural decay may be a dominant loss process that depletes the chemical from the soil. Chemicals with longer half-lives are more persistent in the environment than those with shorter ones.

The basis for most modelling techniques used to estimate the emission of organic contaminants from the soil is Fick's first law and Millington-Quirk's model (1961) of tortuosity. Fick's law is governed by diffusion, which describes contaminant movement due to a concentration gradient within the medium using Equation (2), where J is the chemical flux from the soil into the environment (mg cm^{-2} s^{-1}), D is the chemical-specific effective diffusion coefficient (cm^2 s^{-1}), C is the chemical concentration in soil (mg cm^3), and x is the distance from the contaminant source to the ground surface (cm). The differential term of dC/dx describes the difference in chemical concentration with respect to the difference in depth. The effective diffusion coefficient is determined using Millington-Quirk's model for tortuosity in cases where anisotropy, cementation, and incomplete pore space dispersion are not present:[29]

$$D = d_{air} \frac{P_a^{10/3}}{P_t^2} \tag{3}$$

where d_{air} is the chemical-specific diffusion coefficient in air (cm^2 s^{-1}), P_a is the air-filled porosity (cm^3 cm^{-3}), and P_t is the total porosity (cm^3 cm^{-3}). Literature values for chemical-specific diffusion coefficients are available from several sources[7,19,24] in addition to estimation methods that are available in the literature.[4]

Factors for which site-specific data should be collected include: C, x, P_a, and P_t. The depth to the contaminant source is the distance a chemical molecule must travel to reach the ground surface. Jury *et al.*[30] determined that the soil layer above the compound induces degradation at a rate characterized by the representative half-life of the chemical. Thus, information concerning the depth to the source would indicate the amount of delayed diffusion and degradation the chemical would experience while moving through the uncontaminated layer. Sampling should include analyses to determine the soil type, total porosity, and moisture content typically found on-site. For large sites, more than one sample should be taken to evaluate site-wide soil characteristics, since soil parameters can vary spatially. The US EPA[31] provides tabulated total porosity values for selected soil types. The percentage of water-filled pores (*i.e.* soil moisture content) must then be assumed based on the climatological conditions of the area.

Fate and Transport of Metals in Soil

The fate and transport of metals is influenced by the physicochemical properties of the soil environment. Since most metals do not volatilize or degrade, they are mainly transported in the environment through leaching/erosion, dust resuspension, and surface run-off. When rainfall or surface run-off contacts soil, metals may be leached into groundwater or sloughed downgradient (*i.e.* erosion). Small surface soil particles could be blown off-site as fugitive dust. Once these particles become airborne, metals may deposit onto exposed plants, soil, and surface water bodies in the area. Plants may also accumulate high concentrations of metals through root uptake from contaminated soil.

Unlike organics, the transfer of heavy metals in the environment is mainly influenced by many complex soil factors. Site-specific factors that should be evaluated in any sampling methodology include pH, organic carbon fraction, and redox potential, each of which influences the behaviour of metals in soil. For instance, heavy metals tend to sorb to complex materials, such as clays. Since this reduces their concentration in the soil solution, complexed metals are less likely to be taken up by plant roots or removed via leaching. As with organics, heavy metals tend to absorb onto organic matter, which greatly reduces their mobility within the soil column. The availability of heavy metals is further affected by soil pH. Many, but not all, metals become more mobile under increasingly acidic conditions (*i.e.* a pH of less than 7.0). Several heavy metals form hydrated metal complexes that may become more mobile at higher pH levels (i.e. pH greater than 7.0). The interactions between heavy metals and soil pH is controlled by redox potential, which determines the particular oxidized or reduced form of the metal at a given pH. There are currently several models available (*e.g.* reference 17) that represent the relationship between the pH and redox potential for many common metals and can be used to determine metal speciation in soil.

Case Study: Estimating Indoor and Outdoor Air Concentrations from Subsurface Soil Contamination

An HHRA was performed on an outdoor shopping centre located in Georgia. The purpose of the assessment was to determine vapour emissions from chemicals in soil under both outdoor and indoor air scenarios. The outdoor air scenario included only emissions from unpaved areas. For the indoor air pathway, Jury's Behaviour Assessment Model (BAM) was used to estimate the chemical concentration in indoor air from contaminated soil beneath the building foundation.[32] Air concentrations based on the movement of contaminants from subsurface soil to ambient air through unpaved surfaces was estimated using ASTM's 'Guide for Risk-Based Corrective Action Applied at Petroleum Release Sites'.[33] Although other models are available to model the transport of vapour emissions from subsurface soils, this example is presented to provide a perspective on the types of site-specific data needed to accomplish such fate and transport modelling.

COPCs consisted mostly of metals, polycyclic aromatic hydrocarbons, and volatile organic compounds at relatively low levels. Since metals do not volatilize, this pathway was not quantified for these chemicals. Because of the limited exposure scenarios that were evaluated in this assessment, only data on Henry's Law Constant, effective diffusion rate in air and water, and the soil adsorption coefficient were needed. Chemical-specific parameters used in this assessment are shown in Table 3. Site-specific soil parameters measured included f_{OC}, soil moisture content, total soil porosity, and soil bulk density, as these factors play a pivotal role in determining the persistence of the chemicals in the environment. In the absence of chemical-specific data, air and water diffusion rates were estimated using a technique that involves adjusting diffusion rates known for one chemical to approximate values for other chemicals, based on the assumption that diffusion is inversely related to molecular weight.[8]

Table 3 *Chemical-specific parameters used to model gaseous diffusion through soil*

Chemical	Henry's Law Constant (atm m^3 mol^{-1})[a]	K_{OC} (l kg^{-1})[a]	D_{air} (cm^2 s^{-1})[d]	D_{water} (cm^2 s^{-1})[e]
Acenaphthylene	1.48×10^{-3}	2500	0.063	7.02×10^{-6}
Anthracene	1.02×10^{-3}	14000	0.058	6.487×10^{-6}
Benzene	5.59×10^{-3}	83	0.088[c]	9.8×10^{-6} [c]
Benzo[a]anthracene	1.16×10^{-6}	1380000	0.051[c]	9.0×10^{-6} [c]
Benzo[a]pyrene	1.55×10^{-6}	5500000	0.043[c]	9.0×10^{-6} [c]
Carbon disulfide	1.23×10^{-2}	54	0.104[c]	1×10^{-5} [c]
Chloroform	3.39×10^{-3} [c]	44[b]	0.104[c]	1×10^{-5} [c]
Chrysene	1.05×10^{-6}	200000	0.0515	5.73×10^{-6}
Ethylbenzene	6.43×10^{-3}	1100	0.075	7.8×10^{-6}
Fluoranthene	6.46×10^{-6}	38000	0.0547	6.09×10^{-6}
Fluorene	6.42×10^{-5}	7300	0.0604	6.72×10^{-6}
Methylene chloride	3.19×10^{-3} [c]	8.8[f]	0.101[c]	1.17×10^{-5} [c]
Methyl ethyl ketone	2.74×10^{-5}	4.5	0.806[c]	9.8×10^{-6} [c]
Naphthalene	4.8×10^{-4} [c]	1369[b]	0.059[c]	7.5×10^{-6} [c]
Phenanthrene	1.59×10^{-4}	14000	0.058	6.49×10^{-6}
1,1,2,2-Tetrachloroethane	3.81×10^{-4}	118	0.071[c]	7.9×10^{-6} [c]
Toluene	6.37×10^{-3}	300	0.087[c]	8.6×10^{-6} [c]
1,1,1-Trichloroethane	1.44×10^{-2}	152	0.078[c]	8.8×10^{-6} [c]
Trichloroethene	9.1×10^{-3}	126	0.079[c]	0.1×10^{-6} [c]
Total xylene	6.26×10^{-3} [c]	240[f]	0.076[f]	7.23×10^{-6}

[a] Reference 20; [b] reference 34; [c] reference 19; [d,e] reference 8; [f] reference 7.

Indoor Air Concentrations from Subsurface Soil Contamination.   The indoor air analysis was conducted using Jury *et al.*'s[32] BAM model to estimate the volatilization rate through the soil column coupled with the 'box model' to estimate exposure-point concentrations in air by vapour transport from the soil through cracks in the building foundation. The BAM model is a one-dimensional mathematical model suitable for situations in which a time-dependent vapour emission rate is needed. As prescribed by Henry's Law, the model assumes that the three phases of the chemical (vapour, aqueous, and solid) are in equilibrium. If more than one chemical is present in the soil, the model assumes that there are no interactions between chemicals and that all chemicals are subject to the same hydraulic and soil conditions.

Unlike the more commonly used models, including the Heuristic Model developed by Johnson and Ettinger[35] and ASTM's 'Emergency Standard Guide for Risk-Based Corrective Action Applied at Petroleum Release Sites',[33] the BAM is considered a refined model, since it conserves mass and accounts for the time-varying depletion of contaminants in soil. The model also predicts vapour emission rates by incorporating loss pathways that include chemical transport by volatilization at the soil surface and leaching in the soil column by evapotranspiration. In addition, the BAM model has also been validated in the laboratory and in field studies using soil-incorporated pesticides from the upper root zone.[36]

For this HHRA, the 25-year average emission rate was computed for each COPC. Model requirements for estimating the vapour emission rate included soil porosity, soil bulk density, thickness of the contaminated soil, volumetric water content, total chemical soil concentration, and the diffusion rate of the chemical in air. The emission rate was evaluated by integrating the volatilization flux at the surface from zero to 25 years. This result was then divided by the time period to produce the time-averaged gaseous flux from the soil surface.

Outdoor Air Concentrations from Subsurface Soil Contamination.   Vapour transport through paved portions of the site was assumed to be minimal since the asphalt cover served as an effective barrier. Volatilization rates and air concentrations were predicted using the ambient air equation in reference 33. This approach is similar to the model developed by Farmer *et al.*[37,38] to estimate the emission rate of chemicals from landfills without internal gas generation, assuming that diffusion is the controlling transport mechanism through soil. The model assumptions include constant chemical concentrations in subsurface soils (no loss through biodegradation or leaching); linear equilibrium partitioning within the soil matrix between sorbed, dissolved, and vapour phases; and steady-state vapour- and liquid-phase diffusion through the vadose zone. Emission rate estimates were coupled with the 'box model' to yield quantitative estimates of ambient air concentrations.

Incorporation of site-specific information concerning the soil properties and site conditions including wind speed and the area of unpaved surfaces produced concentrations that were thought to be representative of actual site conditions. For example, the use of a default wind speed of 2 m s^{-1} would imply less wind-blown dilution than actually occurs on-site. The inclusion of site-specific information into the analysis decreased the ambient air concentration by approximately an order of magnitude.

5 Site-Specific Air Data Needs

Air pathway analyses are typically considered when off-site receptors may be potentially exposed to site-related contaminants by inhalation of airborne particulates and vapours. Since particles less than $10\,\mu m$ in diameter may be absorbed by the lung, determining the particulate size distribution is useful. In addition, particulates may deposit onto secondary transport sources (*i.e.* soil, vegetation, and surface water bodies), which could also indirectly affect human exposure. Procedures for assessing indirect and direct impacts associated with exposure to site-related contaminants can be quantified using various air dispersion models, such as those developed by the EPA.[39–41] Site-specific data needed to use air dispersion models include meteorological and site terrain data as well as information on the location of maximum deposition and potentially exposed individuals. For receptors located within close proximity to the source (*i.e.* less than 100 m), air concentrations predicted using air dispersion models are less reliable. Exposures by these receptors should be evaluated using the simple 'box model' approach.

Fugitive Dust Emissions

Fugitive dust emission can be a major concern at contaminated sites that have unpaved and unvegetated areas. The primary mechanisms that influence fugitive dust emissions are wind-blown erosion and soil disturbances caused by human activities (*e.g.* vehicular traffic and excavation activities). Contaminants sorbed to dust may become airborne during remediation activities such as excavation, grading, and truck movement. The loss of soil due to wind erosion is a function of wind velocity, percent of vegetative cover, soil/waste properties such as texture and moisture, and the area of exposed soil. Higher mean annual wind speeds above the threshold velocity, defined as the velocity at which particulates are expected to become airborne, indicate a greater potential for fugitive dust emissions. Dust emissions are not expected to be generated in areas covered by vegetation or concrete. Exposure to fugitive dust is likely to be greater at sites whose soil contains small erodible particulates and has low moisture content (*e.g.* sandy soils). Site-specific data for all of these factors should be evaluated to determine if fugitive dust emission by wind erosion is likely to be a significant transfer mechanism.

Case Study: Estimating Particulate Emissions from Surface Contamination. This section describes a model used to estimate particulate emissions generated by wind erosion for a site near Chicago. Emissions from vehicular traffic were assumed to be negligible relative to impacts from wind erosion, since vehicular traffic was almost exclusively limited to paved areas. A mixture of site-specific and default data were used in the final risk assessment, but had all the default data been used, the concentration of COPCs in respirable particulates would have been over-predicted by a factor of 7.

Vapour Emissions from Point and Area Sources

Gaseous releases into the atmosphere disperse at varying rates in response to numerous site-specific factors controlling their fate and transport. The major concern with airborne hazardous substances is their corresponding ground-level concentration where potential receptors may be exposed. Two types of releases that are typically evaluated in HHRAs are stack releases, such as those from an incinerator, and ground-level area releases typical of landfills.

Emissions from Point Sources.   Factors governing the methodology for estimating maximum ground-level air concentrations from a point source, such as a stack, include plume properties, wind speed, terrain, and atmosphere turbulence. Plume properties include parameters such as plume rise and dispersion of the plume after reaching its maximum height. Plume rise is defined as the height to which a plume rises before bending horizontally and dispersing. Plume rise is a function of the momentum and buoyancy associated with its release from the stack. Stack exit velocity (*i.e.* the velocity at which the plume is released from the source) imparts momentum to the plume. Heated stack gases, especially those from combustion processes, are considerably warmer than the ambient air, giving the plume a lower density, which makes it buoyant in ambient air.[7] The wind controls the direction the plume will move off-site and how dilute the plume concentration will be downwind. As the plume travels downwind, the momentum and buoyancy effects dissipate and atmospheric turbulence becomes the dominant factor determining plume dispersion.

Atmospheric turbulence, or stability, is a classification system used to identify the type of dispersion expected at a site given the wind speed, time of day, and amount of cloud cover. The less stable the atmosphere, the higher the ground-level concentration close to a source. Hence, stability class affects the dispersion pattern and plume profile from the source. Aerodynamic effects due to terrain features such as buildings, trees, and valleys may create localized downwash effects. Complex terrain (*i.e.* locations where plumes intercept hillsides or buildings) are expected to influence the trajectory and diffusion of the plume and cause frequent inversion layers and fumigation effects within the valley basin area. An inversion may create severe conditions by trapping airborne chemicals in the stable layer and allowing contamination to build-up in especially confined topographic areas. Fumigation effects typically draw the chemical to the valley floor during solar heating of the ground. Finally, the emission rate from the stack influences the amount of contaminant expected to impact receptors of concern.

Atmospheric studies conducted in the 1960s, however, revealed that stack emissions from combustion processes are controlled by stack temperature effects more than momentum, and that plume rise is inversely related to the wind speed.

It is difficult to summarize all the factors that influence the maximum ground level concentration from point sources and therefore only major categories of variables are presented for simplicity. They include meteorological considerations, stack parameters, and plume properties. Site-specific data needed include (at a minimum): (1) chemical-specific emission rates, (2) stack height, (3) stack diameter, (4) stack gas temperature, (5) stack exit velocity or flow rate, (6) location of the source with respect to the receptors of interest and the surrounding topography, and (7) detailed description and dimensions of buildings located on-site. Site-specific meteorological data needed include: wind speed, wind direction, and mixing heights.

Vapour Emissions from Area Sources. Emissions from ground-level sources, such as landfills, have minimal momentum and a nominal difference in temperature. Although there are significant differences in molecular weight between air and some released gases, the concentration of the gases is typically too low for density differences to result in buoyancy effects. As a result, plume rise for most ground-level releases at contaminated sites is negligible. For ground-level releases, stable atmospheric conditions typically create the highest concentration along the centreline near the source.

6   Site-Specific Groundwater/Aquifer Data Needs

Groundwater samples need to be collected in a manner that adequately defines the contaminant plume with respect to current and future potential exposure points, including delineation of the horizontal and vertical extent of contamination. Data on aquifer characteristics, including hydraulic conductivity, porosity, and productivity, are also useful for the HHRA. The need to model future groundwater movement to adequately characterize future risks will dictate the specific parameters to be analysed. At a minimum, hydraulic conductivity, porosity, and flow rate should be measured on-site to ensure an accurate prediction of future transport of groundwater off-site.

Hydraulic conductivity (k) is a measure of a soil's ability to transmit water. Hydraulic conductivity ranges from 1×10^{-9} cm s^{-1} for the most impermeable unconsolidated clay to 1×10^{-5} cm s^{-1} for clean gravel. It should be noted that under the best possible environmental conditions, hydraulic conductivity is only measured within an order of magnitude accuracy. In addition, composite soils that consist of a mixture of soil types may vary by orders of magnitude within a confined interval of soil. Other factors that should be measured for mathematical modelling include total porosity and flow rate. The porosity indicates the percentage volume of voids over the total volume in soil and typically ranges from 0.20 to 0.70 cm^3 cm^{-3}. Default values are provided by the EPA[31] for selected types of soil. There are several field methods available for determining groundwater flow rate, and a hydrogeologist should be consulted before tackling the sampling plan. Accurate prediction of groundwater flow and transport within the subsurface soil layer requires that considerable effort is directed towards the accuracy of the sampling methodology before it is implemented. Site-specific parameters mentioned in this section are not meant to form a complete list, but are suggested parameters that may require additional values depending on the models used in the analysis.

Typically, groundwater is considered a drinking water source only if the supply is both potable and pumpable. Hydraulic conductivity of any potentially affected wells should be measured to determine well yield. Total (unfiltered) samples should be obtained to determine potability, as this reflects the level of contaminants to which individuals are most likely to have direct contact. Yield must be at least equivalent to the amount required to provide drinking water for one adult to have sufficient productivity for consideration in the HHRA. To determine if the groundwater is potable, water quality data, such as total dissolved solids, sulfate, and chloride levels should be gathered and compared with applicable criteria.

Volatilization of vapours from groundwater into ambient and indoor air has been a recent concern in many risk assessments. Since groundwater is generally ubiquitous throughout many sites, the potential effects associated with these pathways are substantial. Modelling the transport of groundwater vapours to air is generally based on the concepts described in the Fate and Transport of Organics in Soil Section. The main difference is that the groundwater concentration is converted to a vapour concentration based on the dimensionless Henry's Law Constant. This is the ratio of the vapour fraction to the water fraction for each chemical and may be used to determine the vapour concentration by multiplying by the chemical concentration in groundwater. The site-specific depth of contamination should be set equal to the depth from the ground surface to the water table.

7   Site-Specific Surface Water/Sediment Data Needs

Surface water samples need be collected from on-site or nearby surface water bodies if (1) the water receives or potentially receives discharge from the site, or (2) the water body is used or could be used for recreational purposes (*e.g.* swimming, wading, fishing). Again, total (unfiltered) data should be obtained, as this reflects the level of contaminants to which individuals are most likely to have direct contact. Typically, sediment data are not necessary, since incidental ingestion of and dermal contact with surface water is quantified using total surface water data. Sediment data are needed only if evaluation of direct contact with sediments (*e.g.* while wading) is relevant.

In addition to direct discharge, surface water bodies could be contaminated by groundwater infiltration and surface run-off. For more information on evaluating concentrations due to other anthropogenic sources released into surface water bodies, several models are available. The use of such models, however, can be costly and time-consuming. In general, they are only used if the concentration of contaminants in surface water/sediment is expected to change substantially over time, since it is much cheaper and more accurate to use measured data.

8 Exposure Assessment

Determining if exposure to site-related contaminants may increase the incidence of adverse health effects in exposed populations is an important step in the risk assessment process. The goal of the exposure assessment is to estimate the magnitude, frequency, duration, and route of human exposure to site-related chemicals by completing the following steps: (1) identifying potential human receptors; (2) characterizing pathways of exposure; (3) estimating exposure-point concentrations; and (4) estimating total contaminant intake by potentially exposed individuals for all complete exposure pathways. One objective of the exposure assessment is to facilitate calculation of separate risk estimates for current and future receptors as a function of land use and the measured levels of contaminants in various environmental media.

Potentially Exposed Populations and Exposure Pathways

For risk assessment purposes, an important objective in evaluating the environmental behaviour and fate of various contaminants is predicting the major pathways and extent of human exposure. The relative importance of an exposure pathway depends upon the concentration of COPCs in the relevant medium and the rate of intake by exposed individuals. Potentially exposed individuals are defined in terms of realistic current and future uses of the site. Three types of populations are typically considered in HHRAs: (1) residential, (2) commercial/industrial, and (3) recreational. Residential exposures are typically evaluated whenever individuals currently live on or near a site or future use of the site is commensurate with residential development. Similarly, commercial/industrial exposures should be evaluated for sites that are currently used for commercial/ industrial purposes or are designated as such for future development (*e.g.* areas zoned as industrial/commercial).

Recreational receptors are defined as individuals who do not live on-site but could use the site for recreational purposes. Inclusion of potential recreational use of a site should consider: (1) if access to the site is controlled/limited, (2) the likelihood of non-resident individuals desiring to use the site for recreational purposes, and (3) the distance individuals would have to travel to use the site for recreational purposes. For example, recreational use of a site in an industrial area that is completely fenced is expected to be minimal. To accurately assess recreational exposures, site-specific information on the age distribution of individuals expected to use the site for recreational purposes and information on how recreational activities may change with seasons should be gathered. For example, children are likely to play outdoors less during the winter than in summer, while children are expected to spend more time outdoors in general than adults.

Potential exposure pathways are the routes and media through which contaminants move to reach potential receptors. For an exposure pathway to be considered complete, it must have a source, a mechanism of contaminant release, a retention and transport medium, a point of potential contact, and an exposure route at the contact point. For example, if surface soils are contaminated, direct ingestion of and dermal contact with surface soils are complete exposure pathways, if receptors have direct contact with surface soils. If groundwater is contaminated but current receptors obtain their water from a municipal supply, ingestion, inhalation of organic vapours while showering/bathing, and dermal contact with contaminants while showering/bathing are not complete exposure pathways. Conversely, if the groundwater is potable, and there are no constraints on the use of the supply for domestic purposes in the future, these three exposure pathways would be complete for hypothetical future residents.

Pathways may be eliminated from evaluation if: (1) the exposure resulting from one pathway is lower relative to that from other pathways involving the same medium, (2) the potential magnitude of exposure from a given pathway is low, and (3) the pathway is complete, but the probability of exposure may be low and the associated risks are not high. For example, dermal absorption while showering frequently results in much lower exposures than direct ingestion of groundwater. Similarly, although nearby streams may be affected by site releases, stream productivity is so low that it would be impossible for an individual to consume large amounts of fish from the stream. In this case, although ingestion of fish is a complete exposure pathway, the amount of COPCs ingested via this pathway is expected to be low. Hence, elimination of this pathway is appropriate. Site-specific data on behaviour patterns and the concentration of COPCs in relevant environmental media need to be gathered to support the elimination of complete exposure pathways.

Estimating Chemical Intakes

Chronic daily intakes (CDIs) are typically estimated using the equation:

$$CDI = \frac{C \times CR \times FI \times EF \times ED}{BW \times AT} \qquad (4)$$

where C is the concentration contacted over the exposure period (mg kg^{-1}, mg l^{-1}, mg m^{-3}), CR is the contact rate, or the amount of contaminated medium contacted per unit time (kg day^{-1}, l day^{-1}, m^3 day^{-1}), FI is the fraction of food items consumed from the contaminated area, EF is exposure frequency (days yr^{-1}), ED is exposure duration (yr), BW is average body weight (kg), and AT is averaging time, or the time period over which exposures are averaged (days). Calculated intakes are expressed as the amount of chemical actually taken into the body instead of the amount that is absorbed through the lung or gut once the chemical has been inhaled or ingested. This method of calculating exposures is conservative, as it assumes that 100% of the chemical inhaled or ingested is absorbed by the body. Site-specific data for CR, BW, and AT are not required. Standard (default) contact rates and body weights are typically used. It is difficult to prove that contact rates for individuals who may be exposed to site-related

COPCs differ from standard default values, which typically represent national averages or median values. AT is defined as ED (yr) times 365 days per year for non-carcinogenic COPCs, while AT for carcinogens is specified as 70 years (lifetime) times 365 days per year. Site-specific needs related to determining the concentration of COPCs in various environmental media have been discussed.

Site-specific data for EF and ED are needed as they tend to vary with land use. For example, recreational receptors are expected to spend less time exposed to site-related COPCs than residents, since they do not live on-site. Typically, EF and ED estimates represent the largest source of uncertainty for non-food related exposure pathways. For food chain pathways (*e.g.* ingestion of produce, meat, milk/dairy products, and fish), the largest source of uncertainty stems primarily from estimating the concentration of contaminants in plant or animal tissues in the absence of measured data and the fraction of various food items consumed that originate from or were produced in the contaminated area (FI). To minimize these sources of uncertainty, site-specific estimates of EF, ED, and FI should be obtained. Site-specific surveys represent the least costly and most commonly used method of obtaining site-specific data.[2] A cost-effective strategy for collecting site-specific data on EF, ED, and FI is presented below.

Exposure Assessment Case Study.   Demographic survey statistics for individuals living within or using a large former metals mining site were collected to reduce the uncertainty associated with using standard default exposure values. The Land Use and Demography Survey, as it was called, was used to obtain site-specific information on behavioural patterns and activities so that more accurate exposure and risk estimates could be incorporated in the HHRA. These data were obtained by interviewing individuals representative of the population(s) of interest (*i.e.* adults and children who lived within or used the site). Results of the survey were used to identify more precisely where, under what circumstances, and for how long, potentially exposed individuals may come into contact with contaminated media.

Information on 109 children aged 0 to 6 years was obtained. The results showed that children aged 0 to 6 years spend an average of: 87 h yr^{-1} (3.6 days yr^{-1}) playing on mine/mill waste piles, 870 h yr^{-1} (36.2 days yr^{-1}) outdoors but not on mine/mill waste piles, and 51 h yr^{-1} (2.1 days yr^{-1}) using on-site streams and ponds for recreational purposes. Upper-bound values for these three activities were 123 h yr^{-1} (5.1 days yr^{-1}), 1242 h yr^{-1} (51.7 days yr^{-1}), and 115 h yr^{-1} (4.8 days yr^{-1}), respectively. In an HHRA for a similar site for which site-specific exposure data were not collected, it was assumed (based on best professional judgment) that children who used the site for recreational purposes only would spend 6 hours a day, 2 days a week in April, May, September, and October plus 6 hours a day, 5 days a week in June to August for a total of 23 days a year directly on waste piles. The recreational user was also assumed to spend an additional 15 days a year (2 hours a day, 5 days a week in March to November) outdoors on-site but not directly on waste piles (*e.g.* swimming, fishing). Hence, recreational users were assumed to be exposed to site-related contaminants for 38 days per year. This estimate is slightly lower than the site-specific values of 57 days per year for the upper-bound and 40 days per year for the average exposure scenario obtained from the Land Use and Demography Survey results.

The EPA[1] reports that the average fraction of meat and milk/dairy products that is home-produced is 0.44 and 0.40, respectively, while the worst-case value, which should be used for upper-bound estimates of risk, is 0.75. If residents were assumed to consume these food products twice a day, 350 days a year, then the default average number of meals per year with home-produced meat and milk/dairy products would be 308 (two meals per day, 350 days per year × 0.44 = 308) and 280, respectively. An EF of 350 days per year is used versus 365 days per year to account for the fact that individuals spend 2 weeks away from home.[43] Survey results indicated that children aged 0 to 6 years consume about 15 meals per year that contain beef or game meat from the affected area, which is about 21 times lower than the standard EPA default value. The upper-bound number of meals with meat consumed by children was found to be 36 per year, which is about 15 times lower than the EPA default value of 525 meals per year (two meals per day, 350 days per year × 0.75 = 525). The results for home-produced milk were even more discordant. Children living on-site consumed an average of 0.5 meals per year with milk produced in the area, which is about 560 times lower than the EPA default value of 280 meals per year. The upper-bound number of meals with locally produced milk was 1.3, which is 400 times lower than the EPA default value.

The EPA[1] recommends using average and upper-bound FI values of 0.25 and 0.40, respectively, to model the fraction of home-grown or locally raised garden vegetables consumed daily. The survey results indicated that children age 0 to 6 years consume an average of 52 meals per year that contain fruits and vegetables produced in the area, which is five times lower than the standard EPA default value of 262.5 meals per year, assuming that individuals consume fruits and vegetables three times a day, 350 days a year. The upper-bound number of meals with fruits and vegetables from the area was 80, which is also about five times lower than the EPA default value (420 meals per year). The EPA[44] reports that individuals spend an average of 2.6 hours per day swimming. The survey results showed that children aged 0 to 6 years who live on-site spend an average of 0.14 hours per day participating in a variety of recreational activities (not just swimming) in on-site streams and ponds, which is about 19 times lower than the EPA default value.

In this case, the use of standard EPA default exposure assumptions would have overestimated child average daily intake from ingestion of contaminated food items originating from the affected area by a factor of between 5 and 560. Upper-bound exposure estimates would have been overestimated by a factor of between 5 and 400. Exposure to site-related COPCs in surface water would have been overestimated by a factor of 19. Similar results were obtained for adults (individuals 18 years old and over). These results suggest that EPA default assumptions, which are often based on adult behaviour or consumption patterns, may not be reliable indicators of child exposures and that collecting site-specific data would probably lower exposure and risk estimates. This would, in turn, lower clean-up costs such that for sites with projected clean-up costs of more than $500 000, conducting a site-specific survey may yield substantial cost savings.

9 Conclusions

Collecting site-specific data can ensure a higher level of confidence in final risk estimates by reducing or eliminating the reliance on standard default values or other default data taken from the literature. Many physicochemical properties, such as the K_{OW} value, Henry's law constants, and vapour pressures, can dramatically influence the fate and transport of site-related chemicals. Since there is tremendous variability for some of these parameters (see Table 2), the collection of site-specific data for these properties is recommended. Site-specific data are needed to evaluate the fate and transport of chemicals in air, water, and soil and to more accurately quantify potential risks associated with human exposure to chemicals in these media (Table 4).

The advantage of using site-specific data to model outdoor air concentrations from subsurface soil contamination has been illustrated through the presentation of case studies. Inclusion of site-specific data into the model yielded ambient air concentrations one order of magnitude lower than those predicted using standard default assumptions.

Estimating emissions from point sources (*e.g.* stacks) can be a highly uncertain exercise due to the complicated air dispersion modelling that must be performed. Collection of site-specific information on: (1) meteorological conditions (including annual average wind speed and precipitation data); (2) stack characteristics (including stack height, diameter, and exit velocity); (3) terrain and topography; and (4) the location of individuals relative to the source, are essential, at a minimum, for reducing the level of uncertainty. For groundwater, site-specific data such as hydraulic conductivity (a measure of how much water can be pumped) and potability (as determined by measuring various water quality parameters) are necessary to determine if potential receptors are likely to consume contaminated groundwater.

Typically, exposure frequency and duration estimates represent the largest source of uncertainty for non-food-related exposure pathways, while the largest source of uncertainty for food chain pathways stems primarily from estimating the fraction of various food items consumed that originate in the contaminated area. Site-specific data on exposure frequency and duration are often needed, as these parameters tend to vary with age and land use. The results of a case study show that the use of standard default exposure assumptions would have overestimated child exposures from ingestion of contaminated food items by a factor of 5 to 560. These results suggest that default assumptions, which are often based on adult behaviour or consumption patterns, may not be reliable indicators of child exposures.

Table 4   *Summary of site-specific parameters and their role in human health risk assessment*

Parameter	Relevance to fate and transport modelling	Comments
Water solubility $(mg\ l^{-1})$	Determines extent to which chemical will dissolve in water	Highly soluble chemicals tend to leach rapidly from soil to water
K_{ow} value (Dimensionless)	Predicts extent to which chemical will sorb to soil and other organic matter	Chemicals with high K_{ow} values tend to sorb to soil, sediment, and accumulate in the food chain
Vapour pressure (mmHg)	Determines the extent to which chemical will partition into air	Chemicals with high vapour pressures tend to partition into air and have a low affinity for soil, inhalation being a primary exposure pathway
Henry's law constant (atm $m^3\ mol^{-1}$)	Determines the fraction of the chemical that will partition into the gaseous and liquid phases	Chemicals with high Henry's law constant tend to partition into air and have a low affinity for water. Highly sensitive to temperature changes
Bioconcentration and biotransfer factors (BCFs and BTFs)	Predicts the extent to which chemicals will accumulate in living organisms	Larger BCFs and BTFs indicate that the chemical is likely to accumulate in food items consumed by humans
Organic carbon partition coefficient (K_{OC})	Measures relative sorption potential for organic chemicals	Higher K_{OC} values indicate greater tendency to sorb to soil
Chemical flux from soil (J, in mg $cm^{-2}\ s^{-1}$)	Basis for modelling organic emissions from soil	
Chemical-specific effective diffusion coefficient (D, in $cm^2\ s^{-1}$)	Basis for modelling organic emissions from soil	
Distance from contaminant source to ground surface (x, in cm)	Affects the rate at which chemicals will diffuse through soil	
Air-filled soil porosity (P_a in $cm^3\ cm^{-3}$)	Affects the rate at which chemicals will diffuse through soil	
Total soil porosity (P_t, in $cm^3\ cm^{-3}$)	Affects the rate at which chemicals will diffuse through soil	
Chemical-specific diffusion coefficient in air (d_{air}, in $cm^2\ s^{-1}$)	Determines the rate at which a chemical will diffuse through air	
Soil density, type, and moisture content	Affects the rate at which chemicals will diffuse through soil	

Table 4 *Continued*

Parameter	Relevance to fate and transport modelling	Comments
Wind speed (m s^{-1}), percent vegetative cover, as well as soil type and moisture content	Affect fugitive dust emissions and inhalation of resuspended dust	
Particle size distribution of soil particulates	Affects fraction of soil particles that may be inhaled	Particles <10 μm in size are considered respirable
Stack exit velocity, diameter, and temperature	Needed to conduct air dispersion modelling	Greater velocity equals greater dispersion of chemicals. Higher release temperatures make plume more buoyant
Wind speed, wind direction, and mixing height	Needed to conduct air dispersion modelling	Affects direction and speed of chemical dispersion
Atmospheric turbulence (stability)	Needed to conduct air dispersion modelling	A less stable atmosphere yields higher ground-level air concentrations near the source
Terrain specifications	Needed to conduct air dispersion modelling	The presence of hills or buildings can impede plume dispersion
Hydraulic conductivity (k, in cm s^{-1})	Measure of a soil's ability to transmit water (well yield)	Groundwater is classified as a drinking water source only if pumpable and potable
Water quality data	Used to evaluate the potability of the water supply	Groundwater is classified as a drinking water source only if pumpable and potable

References

1 US Environmental Protection Agency (EPA), 'Risk Assessment Guidance for Superfund, Volume I: Human Health Evaluation Manual (Part A)', EPA/504/1-89/002, Office of Solid Waste and Emergency Response, EPA, Washington, DC, 1989.

2 US Environmental Protection Agency (EPA), 'Guidelines for Exposure Assessment, Federal Register, 57(14)', EPA, Washington, DC, 1992, pp. 22888–22938.

3 R. E. Menzer and J. O. Nelson, 'Water and Soil Pollutants', in 'Toxicology', eds. J. Doull, C. D. Klaassen, and M. D. Amdur, Macmillan, London, 1980.

4 W. J. Lyman, W. F. Reehl, and D. H. Rosenblatt, 'Handbook of Chemical Property Estimation Methods', McGraw-Hill, New York, 1982.

5 G. C. Briggs, 'Theoretical and experimental relationships between soil adsorption, octanol–water partition coefficients, water solubilities, bioconcentration factors, and the parachor', *J. Agric. Food Chem.*, 1981, **29**, 1050–1059.

6 A. J. Dobbs and M. R. Cull, 'Volatization of chemicals – relative loss rates and the estimation of vapor pressures', *Environ. Pollut. (Ser. B)*, 1982, **3**, 289–298.

7 M. D. LaGrega, P. L. Buckingham, and J. C. Evans, 'Hazardous Waste Management', McGraw-Hill, New York, 1994.

8 R. P. Schwarzenbach, P. M. Gschwend, and D. M. Imboden, 'Environmental Organic Chemistry', Wiley-Interscience, New York, 1993.

9 C. C. Travis, H. A. Hattemer-Frey, and A. D. Arms, 'Relationship between dietary intake of organics and their concentrations in human adipose tissue and breast milk', *Arch. Environ. Toxicol. Chem.*, 1987, **17**, 473–478.

10 C. T. Chiou, V. H. Freed, D. W. Schmedding, and R. L. Kohnert, 'Partition coefficient and bioaccumulation of selected organic chemicals', *Environ. Sci. Technol.*, 1977, **11**(5), 475–478.

11 S. F. J. Chou and R. A. Griffin, 'Solubility and soil mobility of polychlorinated biphenyls', in 'PCBs in the Environment, Volume I', ed. J. S. Waid, CRC Press, Boca Raton, FL, 1986, pp. 101–120.

12 H. Geyer, P. Sheehan, D. Kotzias, D. Freitag, and F. Korte, 'Prediction of ecotoxicological behavior of chemicals: Relationship between physico-chemical properties and bioaccumulation of organic chemicals in the mussel *Mytilus edulis*', *Chemosphere*, 1982, **11**(11), 1121–1134.

13 H. J. Geyer, I. Scheunert, and F. Korte, 'Correlation between the bioconcentration potential of organic environmental chemicals in humans and their *n*-octanol/water partition coefficients', *Chemosphere*, 1987, **16**(1), 239–252.

14 E. E. Kenaga, 'Correlations of bioconcentration factors in aquatic and terrestrial organisms with their physical and chemical properties', *Environ. Sci. Technol.*, 1980, **14**, 553–556.

15 E. E. Kenaga and C. A. I. Goring, 'Relationship between water solubility, sorption, octanol–water partitioning, and concentration of chemicals in biota', in 'Aquatic Toxicology', eds. J. E. Eaton, P. R. Parrish, and A. C. Hendricks, American Society for Testing Materials, Philadelphia, STP 707, 1980.

16 C. C. Travis and A. D. Arms, 'Bioconcentration of organics in beef, milk, and vegetation', *Environ. Sci. Technol.*, 1988, **22**, 271–274.

17 J. Dragun, 'The fate of hazardous materials in soil (what every geologist and hydrogeologist should know). Part 2', *Hazard. Mater. Control*, 1988, **1**(3), 41–65.

18 Arthur D. Little, Inc. (ADL), 'Installation Restoration Program Toxicology Guide', ADL, Cambridge, MA, 1987.

19 US Environmental Protection Agency (EPA), 'Hazardous Waste Treatment, Storage, and Disposal Facilities (TSDF) – Air Emission Models', EPA-450/3-87-026, Office of Air and Radiation, EPA, Research Triangle Park, NC, 1989.

20 US Environmental Protection Agency (EPA), 'Superfund Public Health Evaluation Manual', EPA/540/1-86/060, Office of Emergency and Remedial Response, EPA, Washington, DC, 1986.

21 D. Mackay, S. Paterson, and W. H. Schroeder, 'Model describing the rates of transfer of organic chemicals between atmosphere and water', *Environ. Sci. Technol.*, 1986, **20**(8), 810–815.

22 T. E. McKone, 'The use of environmental health-risk analysis for managing toxic substances', presented at the 78th Annual Meeting of the Air Pollution Control Association, Detroit, MI, 1985.

23 P. J. Shea, J. B. Weber, and M. R. Overcash, 'Uptake and phytotoxicity of di-*n*-butyl phthalate on corn (*Zea mays*)', *Bull. Environ. Contam. Toxicol.*, 1982, **29**, 153–158.

24 T. T. Shen, 'Estimating hazardous air emissions from disposal sites', 1981, *Pollut. Eng.*, **13**(8), 31–34.

25 G. A. Lugg, 'Diffusion coefficients of some organic and other vapors in air', *Anal. Chem.*, 1968, **40**(7), 1072–1077.

26 US Environmental Protection Agency (EPA), 'Estimating Exposure to Dioxin-Like Compounds', Vol. II, EPA/600/6-88/005Cb, Office of Research and Development, EPA, Washington, DC, 1994.

27 S. W. Karickhoff, 'Semi-empirical estimation of sorption of hydrophobic pollutants on natural sediments and soils', *Chemosphere*, 1981, **19**(8), 833–864.

28 D. Mackay, 'Correlation of bioconcentration factors', *Environ. Sci. Technol.*, 1982, **16**, 274–278.

29 R. J. Millington and J. M. Quirk, 'Permeability of porous solids', 1961, *Trans. Faraday Soc.*, **57**, 1200–1207.

30 W. A. Jury, D. Russo, G. Streile, and H. El Abd, 'Evaluation of volatilization by organic chemicals residing below the soil surface', 1990, **26**(1), 13–20.

31 US Environmental Protection Agency (EPA), 'Water Quality Assessment: A Screening Procedure for Toxic and Conventional Pollutants in Surface and Groundwater – Part II (Revised 1985)', EPA/600/6-85/002b, Environmental Research Laboratory, EPA, Athens, GA, 1985.

32 W. A. Jury, W. F. Spencer, and W. J. Farmer, 'Behavior assessment model for trace organics in soil: I. Model description', *J. Environ. Qual.*, 1983, **12**(4), 558–564.

33 American Society for Testing and Materials (ASTM), 'Emergency Standard Guide for Risk-Based Corrective Action Applied at Petroleum Release Sites', ES 38-94, ASTM, Philadelphia, 1994.

34 J. H. Montgomery and L. Welkom, 'Groundwater Chemicals Desk Reference', Lewis Publishing, Chelsea, MI, 1990.

35 P. C. Johnson and R. A. Ettinger, 'Heuristic model for predicting the intrusion rate of contaminant vapors into buildings', *Environ. Sci. Technol.*, 1991, **25**(8), 1445–1452.

36 W. A. Jury, W. F. Spencer, and W. J. Farmer, 'Behavior assessment model for trace organics in soil: IV. Review of experimental evidence', *J. Environ. Qual.*, 1984, **13**(4), 580–586.

37 W. J. Farmer, M. S. Yang, J. Letey, W. F. Spencer, and M. H. Roulier, in 'Land Disposal of Hazardous Waste', Proceedings of the Fourth Annual Research Symposium, Report No. 631/9-78-016, US Environmental Protection Agency, Washington, DC, 1978, pp. 182–190.

38 W. J. Farmer, M. S. Yang, J. Letey and W. Spencer, 'Land Disposal of Hexachlorobenzene Wastes: Controlling Vapor Movement in Soil', EPA-600/2-80/119, Municipal Environment Research Laboratory, Cincinnati, OH, 1980.

39 US Environmental Protection Agency (EPA), 'Guidelines on Air Quality Models (Revised)', Draft Report EPA-450/2-87-027R, Office of Air Quality Planning and Standards, EPA, Research Triangle Park, NC, 1984.

40 US Environmental Protection Agency (EPA), 'Industrial Source Complex (ISC) Dispersion Model User's Guide – Second Edition (Revised)', EPA-450/4-88-002a, Office of Air Quality Planning and Standards, EPA, Research Triangle Park, NC, 1987.

41 US Environmental Protection Agency (EPA), 'Supplement to a Guideline on Air Quality Models (Revised)', Draft Report EPA-450/4-78-027R, Office of Air Quality Planning and Standards, EPA, Research Triangle Park, NC, 1987.

42 US Environmental Protection Agency (EPA), 'Rapid Assessment of Exposure to

Particular Emissions from Surface Contamination Sites', EPA/800/8-85/002, Office of Health and Environmental Assessment, EPA, Washington, DC, 1985.

43 US Environmental Protection Agency (EPA), 'Risk Assessment Guidance for Superfund, Volume 1, Human Health Evaluation Manual, Supplemental Guidance Standard Exposure Factors', Draft Final, OSWER Directive 9285.6-03, Office of Solid Waste and Emergency Response, EPA, Washington, DC, 1991.

44 US Environmental Protection Agency (EPA), 'Superfund Exposure Assessment Manual', EPA 540/1-88-001, Office of Remedial Response, EPA, Washington, DC, 1988.

Chapter 12

Information Sources Covering the Environmental Impact of Chemicals

By Mike Hannant[1] and Paula Owen[2]

[1]THE ROYAL SOCIETY OF CHEMISTRY, THOMAS GRAHAM HOUSE, THE SCIENCE PARK, MILTON ROAD, CAMBRIDGE CB4 4WF, UK
[2]THE BRITISH LIBRARY, ENVIRONMENTAL INFORMATION SERVICE, 25 SOUTHAMPTON BUILDINGS, LONDON WC2A 1AW, UK

1 Introduction

It is important to remember that when undertaking any research or study of the environmental impact of chemicals, there may be a sizeable amount of relevant data already available in the literature both in printed and electronic form. Therefore it may be the case that no original research is required. However, it must also be understood that information on the environmental impact of many chemicals does not exist in the literature. The absence of such data does not necessarily mean that it does not exist, it merely means that the information is not in the public domain. Many companies and government agencies hold environmental data on chemicals and class it as 'confidential'. This, of course, does not help the enquirer in their quest for that information.

This chapter sets out to describe where the information resides in the public domain, how it is published now (describing the key sources of environmental data), and how new technology is making it easier to access information and share knowledge on environmental issues. It also describes the environmental services offered by the British Library through its 'Environmental Information Service' and the environmental information offered by The Royal Society of Chemistry.

2 How Information is Published

We are currently experiencing a revolution in the way information is published and distributed. Nowadays, we find more and more information coming to us in some sort of electronic format rather than the traditional printed source. Indeed there is talk of journals being produced exclusively in electronic format in the future or new journals being made available on the Internet (through a subscription service). Environmental science is an interesting area to study when looking at information sources. As a result of its relatively recent emergence as a science in its own right there are no preconceived 'hard and fast' rules about how the information should be published, unlike a discipline such as chemistry for example.

It can be said that environmental information, more than any other scientific discipline, is available to the enquirer in a plethora of formats, for example:

- Traditional printed sources (primary, secondary, and tertiary)
- Grey literature (primary and secondary)
- Electronic sources (mainly secondary)
- Internet/WWW (World Wide Web) sources (mixture of primary and secondary)
- Personal contact

By *traditional sources*, one simply means printed or hard copy sources of the information. There are five main types of traditional source, namely: books, journals, patents, directories, and secondary sources such as abstracting or indexing journals. This type of information is usually the most extensively advertised and easy to access through book shops, publishers, or libraries.

Grey literature is very important in the environmental information field as a significant proportion of important facts and data are reported in this format that cannot be found elsewhere. The problem with grey literature is that it is notoriously difficult to locate as there is no one standard procedure for classifying it. One attempt to organize and collate information on grey literature comes from the database SIGLE (System for Information on Grey Literature in Europe)[1] produced by the British Library (see also Section 3).

Electronic sources consist of CD-ROM and on-line sources. There are (usually) no new sources of information in this type of medium. Information is taken from the printed primary and secondary sources and converted into electronic format with searchable indexes. There are a number of important sources for environmental searchers, databases such as Enviroline,[2] Environmental Bibliography,[3] Pollution Abstracts,[4] and Aqualine.[5] Even the more general databases such as Chemical Abstracts,[6] Compendex (engineering database),[7] and CAB Abstracts[8] offer a wealth of information on environmental issues.

There are, databases, however, that do carry primary information and data; these are known as databanks. ECDIN (Environmental Chemicals Data Information Network)[9] and HSDB (Hazardous Substances Data Bank)[10] are two examples of databanks which supply the enquirer with data such as that concerning human and animal health, LD_{50} (dose that is lethal to 50% of test subjects) toxicity levels or the environmental fate of chemicals once they have been released into the environment. These databases supply a compendium of data that would be difficult and more time-consuming to find from other sources.

Internet/WWW sources. The Internet has been separated from the other electronic media as there is a succinct difference in the type of information available from Internet or WWW sources.

As mentioned above, the majority of electronic sources act as secondary sources of information, simply reporting on details of primary source information.

The Internet is different in that the material now available almost matches the range of material available in other formats. For example, there are electronic journals; searchable databases and library catalogues; e-mail discussion groups and bulletin boards; up-to-date environmental texts and documents; technical reports and conference proceedings; computer software, images, and sound archives that can be downloaded. This allows the environmental enquirer a larger range of sources than any other single medium.

In addition, most of the active environmental organizations and government agencies now have their own home pages on the Internet. Institutions ranging from Greenpeace to the US Environmental Protection Agency have information about themselves and other related organizations that is easily accessible through their home page. Some pages even have searchable databanks available and hence one is able to explore individual topics. Sometimes even images can be accessed and downloaded without copyright restrictions.

One of the most useful functions of the Internet, however, is its ability to put people in touch with experts and other interested parties in most environmental fields around the world. Through the use of e-mail it is possible to join pertinent groups and bulletin boards that focus on your topic of interest. Within the groups one will find experts from academia, commerce, and industry who specialize in the area. However, it is important to be very selective in choosing discussion groups: some of them are very active and produce copious amounts of correspondence that may not be of any relevance.

Personal contact: Sometimes this source is the most useful and timely method of retrieving information. Instead of wasting valuable time searching around for the right book or journal that could take days, one telephone call to the relevant person could answer your query or point you in the direction of the most useful sources of information. The only disadvantage to this source of information is that a good contact list takes time to build up and also people move on and new contacts have to be made.

3   Problems with the Availability of Environmental Information

Environmental information can be very difficult to get hold of as it is so varied in scope and format. A search can also become very involved if one is looking for an impartial overview of a particular subject rather than one fixed aspect of it. This difficulty results from the manipulation of environmental data and information by various agencies or pressure groups in order to better suit their aims and objectives.

Some of the problems encountered when trying to locate environmental information are outlined below:

Split discipline: Environmental science has only been considered a science in its own right for the last 10 to 20 years. Before that time environmental science was simply a subset of the other, longer established, sciences such as chemistry, biology, ecology, *etc*. The interconnection with other sciences makes searching for solutions to environmental issues a multidisciplinary procedure. It can be said that if you are searching for chemical information, for example, you would need to consult a chemist. However, if you are searching for environmental information you need to consult not only an 'environmental specialist', but a chemist, an engineer, an ecologist, a medical doctor, and so on.

Subjectivity: As a result of the sometimes emotive and often controversial nature of environmental issues, subjectivity becomes more of an issue in this area than any other. Care must be taken to examine all sides of an environmental problem, as, for example, a planning issue from an environmentalist's point of view would take a very different tone to that of a commercial developer's report on the same problem even if they both purport to be 'environmental reports'.

Cost of gathering data: Information gathering in this area can be very costly. For example, field studies and environmental auditing all involve expensive processes and experts to undertake the work. This can be particularly important in measuring the effects of air, soil, and water pollution at different localities. As a result the data are sometimes not available for public consumption at the time when the information is needed.

Restrictions on access: Organizations with special collections may restrict access or charge for their information. A company, for example, may be unwilling to release information about emissions. Even some public bodies that are charged with collecting data may not necessarily be prepared to release it, even to *bona fide* enquirers. If the data are made available it may be at such a charge that would discourage frequent requests. There has been much debate over such policies and many public bodies are very reluctant to charge for data that are badly needed by countries and organizations that have limited financial means. Unfortunately, there is increasing pressure on such suppliers of information to recover their costs, especially if their own funding from public money is diminishing, and so an increase in charging for information is therefore likely.

Grey literature: Grey literature abounds in the environmental area and the main drawback with it, as mentioned earlier, is that is very difficult to trace. Grey literature consists of publications that are outside the easily recordable categories such as books, encyclopaedias, and journals. Typical grey literature, for example, could be conference proceedings, reports, newsletters, bulletins, theses, and pamphlets, all of which often contain important information that is not found elsewhere, and is often not recorded in the standard information sources.

Misleading information: Care has always to be taken to ensure that one does not inadvertently provide misleading information. This could be caused by not providing all the information that exists on a particular subject that is readily available or by giving the impression that there is no more information on a subject when in reality there is. It can also be misleading to present information in such a way that it is misunderstood, or used incorrectly.

Validity and interpretation of data/information: An information provider needs to check the validity of information by giving the source of that information and by providing as much information as possible to enable the user to decide how credible the information is. Interpretation is a different story. Users often want the information they receive to be interpreted. Information providers should be careful not to do this unless they have a particular knowledge of the environmental issues involved.

To turn to the positive side, the rapid growth of interest in the environment has lead to an equally rapid growth in the provision of environmental information, not only of a scientific and technical nature, but also concerned with legislative, regulatory, and commercial activities. This growth in provision is apparent not only in printed and electronic sources but also in the number of organizations offering environmental information services. Many information sources that cover the environment among other areas have become 'greener' as they increase their coverage of environmental information. This can only be an advantage for the environmental enquirer.

4   Where Libraries Can Help: The British Library

Most libraries will nowadays include an environmental section, however small, within their collection. This may consist of one or two major works on environmental issues and a number of leaflets on local projects, or it may have a dedicated environmental section whose book titles run into hundreds. Within the British Library there is a large environmental selection of titles and journals, and although the British Library is not a lending library it is possible to photocopy most articles or passages (subject to copyright laws) or to borrow items through interlibrary loan from the Boston Spa division of the British Library. In addition the British Library operates the Environmental Information Service which aims to answer the majority of the environmental enquirer's needs.

The Role of the Environmental Information Service

The Environmental Information Service (EIS) was originally founded in 1989, with initial co-operation and assistance of the Confederation of British Industries, to provide environmental information to industry and other commercial concerns. Since then the remit of EIS has expanded to service all areas of society, this expansion being a result of the demand for environmental information dramatically increasing in these sectors.

The British Library's EIS offers a number of services within the environmental field. Its remit now includes all sectors of the community from the general public and students through to industry, consultancies, and government offices.

Environmental enquiry point: This is basically a free enquiry service for immediate answers to enquiries about names and addresses of environmental organizations, suppliers, and specialists. In addition checks can be run on literature held by the library on environmental issues. One of the aims of EIS is to be a 'first stop' shop where people come with specific enquiries and EIS then attempts to put them in touch with the most relevant person or organization. However, if an enquiry turns out to be involved and/or will take up a large percentage of staff time, then the enquirer will be informed that a charge will have to be made to help recover basic staffing costs.

On-line priced enquiry service: Through *STM Search* the EIS provides expert literature searchers covering all areas of the environment as well as other areas such as science, technology, and medicine. With skilled on-line searchers and access to over 400 databases the service provides quick and cost-effective access to information. The initial consultation is free and a quotation is then given for the search itself.

All of the major environmental databases are accessible through *STM Search*. For example, Enviroline, Environmental Bibliography, Aqualine, Pollution Abstracts, ECDIN, Oceanic Abstracts,[11] Toxline,[12] and CAB abstracts.

Environmental information courses: The environmental course 'Source of Environmental Information' is run approximately three times each year and offers participants a day-long lecture course on the different aspects of environmental information. Included in the day is a series of lectures from British Library staff on various sources of information; an Internet lecture and demonstration of environmental sources available; a series of case study lectures from external speakers from various environmental areas; and a number of demonstrations of databases and on-line systems. In addition, participants are given a demonstration of the new EIS internet home page (see below).

Environmental publications: The British Library also produces a number of publications on environmental issues. Recent titles on the environment include: 'Environmental Information – a guide to sources (second edn), BS7750 – what it means to you', 'Business and Environmental Accountability and Environmental Auditing'. In addition, a short guide to finding environmental information is also available.

Environment Information Service home page: The EIS home page is the most recent addition to the range of services offered. Over the last 6 months, EIS has been making more and more use of the Internet for answering free enquiries; presently 20% of free enquiries are answered this way. As a natural progression from its use of the Internet, it was thought appropriate that EIS should develop its own home page to help Internet users locate relevant environmental information for themselves.

EIS staff are presently working on the following items to add to existing EIS pages: Briefing sheets: A series of information sheets on popular, often requested environmental issues.

Bibliographic lists: Lists of references relating to the environment (see appendix A). Information on other sources: EIS has used many other organizations' home pages and databases when researching environmental topics. Here lists are supplied of other organizations by alphabetical order. Most organizations are accompanied by a paragraph briefly listing what their pages contain and what their main interests are (see Appendix B).

Hypertext links to useful organizations: Through hypertext links an enquirer can be connected to any of the organizations listed in the home page. It is not necessary to know the internet address of the organization at this stage as the hypertext link will take you straight through to the relevant home page.

Diary and forthcoming environmental events: There will be a diary for EIS events and seminars along with dates for any other organization's events of environmental interest.

5 Environmental Information from The Royal Society of Chemistry

The Royal Society of Chemistry (RSC) is charged under its Royal Charter, which sets out the objectives of the RSC, with "the advancement of the science of chemistry and its applications and the maintenance of high standards of competence and integrity among practising chemists" and also "to foster and encourage the growth of application of such chemical science by dissemination of chemical knowledge". To fulfil this latter part of the Charter the RSC publishes a range of information products in both printed and electronic form covering various aspects of information on chemicals and chemistry, including the environmental effects of chemicals.

The RSC also operates a library and information centre in central London which specializes in the provision of chemistry-related information to members, corporate subscribers, and others on a fee-paying basis.

The RSC publishes environmental information in a number of forms. These can be effectively subdivided into:

- Journals
- Current awareness services
- Books
- Data compilations

and these are available in a variety of forms: printed, on-line and CD-ROM.

Journals

Issues in Environmental Science and Technology.[13] In response to the rapid growth of interest in the environment and the acute need for concise, authoritative, and up-to-date reviews of topical issues, the RSC, in 1994, launched this new review journal. Two issues of the journal are published each year, and each issue is devoted to a specific environmental topic. The reviews are written by world experts in specialized fields of environmental science. The series presents a multidisciplinary approach to pollution and environmental science, and focuses on broader issues such as economic, legal, and political considerations.

The titles so far produced in the series are:

- Mining and its Environmental Impact (1994)
- Waste Incineration and the Environment (1994)
- Waste Treatment and Disposal (1995)
- Volatile Organic Compounds in the Environment (1995)
- Agricultural Chemicals and the Environment (1996)
- Environmental Impact of Chlorine Chemistry (1996)

Current Awareness Services

The RSC produces two current awareness services covering the environmental impact of chemicals:

- Chemical Safety NewsBase
- Environmental Chemistry, Health and Safety

Chemical Safety NewsBase.[14]   Chemical Safety NewsBase (CSNB) is an on-line database containing bibliographic information, abstracts, and indexing derived from journal articles, reviews and news items, books, and other printed publications (including legislation, Standards, and data sheets). The database covers a wide range of information on the hazards of working with chemicals in the working environment. In the environmental area it contains information on spillages, disposal, and accidental release of chemicals (see Table 1).

The database is available via KR DIALOG, Data-Star, STN International, and Questel-Orbit, and contains over 37 000 records from 1981 to date. The database is updated monthly with approximately 250 new items.

Information from the database is also available in monthly current awareness publications:

- Laboratory Hazards Bulletin
- Chemical Hazards in Industry

These subsets of Chemical Safety NewsBase provide information specifically for laboratories and chemical hazards within the chemical industry.

Table 1   *Sample record from CSNB online via Data-Star*

AN	HAZ1504000815 9504.
TI	Draft guidance issued to environment agency on sustainable development.
SO	Jan 1995.
YR	95.
PT	q: press release.
LG	EN.
SC	07: reproductive hazards.
DE	environment; pollution-control; Environment-Agency; legislation-UK.
CO	DoE 2 Marsham Street, London SW1P 3EB UK.
AB	A draft outline showing the scope of guidance ministers intend to give to the Environment Agency on their contribution to sustainable development has been issued. On the functions and purposes of the Environment Agency in the field of pollution control the guidance includes compiling information relating to pollution and following relevant developments in technology and techniques. Pollution control functions apply in integrated pollution control; radioactive substances; waste regulation; water discharge consents, water protection zones, nitrate-sensitive areas and anti-pollution works; contaminated land and abandoned mines.

END OF DOCUMENT

Environmental Chemistry, Health and Safety.[15] Environmental Chemistry, Health and Safety (ECH&S) is a CD-ROM product made available through KR OnDisc. It provides comprehensive scientific and technical information on the effects of chemicals deemed to cause actual or potential problems to humans and/or the environment. Environmental information from the scientific standpoint as well as the legal aspect is quoted.

The product contains selected abstracts from a number of the RSC's databases including:

- Analytical Abstracts
- Chemical Business NewsBase
- Chemical Safety NewsBase
- Chemical Engineering and Biotechnology Abstracts

ECH&S contains 130 000 items (as of June 1995) and is updated every 3 months with an extra 3000 items (see Table 2).

Books

The RSC has a very active and prolific book commissioning department that publishes books in a wide range of areas of chemistry. There are a number of texts that fall within the area of the environment and are very useful in keeping abreast of current thinking and current practices. Books of particular note are:

- Risk Management of Chemicals (1992).[16] An authoritative work that reviews the current status of risks entailed in the manufacture, handling, use, and disposal of chemicals and suggests future action for the protection of both the workplace and the natural environment.

Table 2   *Sample record from KR OnDisc ECH&S*

---

2 of 19 Complete Record
   3036457
   Title: EPA resources via Internet
   Journal: Lab. Saf. Environ. Manage.   Vol 2   Iss 6   Pg 3
   Publication Date: NOV–DEC 1994
   Language: English
   Document Type: Journal

   Abstract: The EPA opened a pilot one directional ListServe network in Oct 1994 to distribute select Federal Register documents automatically on the day of publication. Documents can be obtained directly from the US Government Printing Office database. Files available include: air and radiation; pesticides; office of pollution prevention and toxic substances documents (excluding the Toxic Release Inventory); hazardous and solid waste; water; and press releases. Further information can be obtained from John A. Richards tel. 001-202 260-225.

- Organic Substances in Soil and Water (1993).[17] This book documents the latest research on the subject and reviews the function, properties, and structure of organic substances in influencing the behaviour and fate of chemical contaminants in soil, surface water, and groundwater systems.
- Integrated Pollution Control (1994).[18] By presenting a diversity of case studies, this publication shows how environmental legislation now affects industry.
- Carbon Dioxide Chemistry: Environmental Issues (1994).[19] This book covers carbon dioxide chemistry with a view to understanding its effects on the environment.
- The Chemical Industry – Friend to the Environment? (1992).[20] This book gives a state-of-the-art account of the feelings and approaches of leading companies to environmental issues and gives examples for other companies to emulate.

Data Compilations

The Dictionary of Substances and their Effects.[21]   The Dictionary of Substances and their Effects (DOSE) is a seven-volume work covering basic physico-chemical properties and hazard information on over 4000 chemicals. The chemicals appearing in DOSE were selected because they are known to have an adverse effect on certain living organisms. DOSE contains, in summarized but well referenced format, both physical/chemical profiles and mammalian and ecotoxicological data. The information in DOSE has been collected and arranged in such a way to provide users with all the relevant data necessary to identify the hazards associated with a chemical. The extensive referencing of each entry allows readers to obtain more detailed information on individual sections of specific interest. The data in DOSE is organized under the data headings summarized below:

Identifiers
Chemical name
Structure
CAS Registry Number
Synonyms
Molecular formula
Molecular weight
Uses
Occurrence

Physical properties
Melting point
Boiling point
Flash point
Specific gravity*
Partition coefficient*

Volatility*
Solubility*

Occupational exposure
Limit values
UN number
HAZCHEM code
Conveyance classification
Supply classification
Risk phrases
Safety phrases

Ecotoxicity
Fish toxicity*
Invertebrate toxicity*
Bioaccumulation*
Effects on non-target species*

Environmental fate
Nitrification inhibition*
Carbonaceous inhibition*
Anaerobic effects*
Degradation studies*
Abiotic removal*
Absorption*

Mammalian and avian toxicity
Acute data
Sub-acute data
Carcinogenicity and long-term effects
Teratogenicity and reproductive effects
Metabolism and pharmacokinetics
Irritancy
Sensitization
Genotoxicity
Any other adverse effects to man

Any other adverse effects

Legislation

Any other comments

References

*These data headings could be required in an environmental assessment of a particular chemical.

Ecotoxicity. In the ecotoxicity section, information is presented on the effects of chemicals on various ecosystems. Results of studies carried out on aquatic species, primarily fish (LC_{50}, with duration of exposure are quoted for two species of freshwater and one marine species if available) and invertebrates [LC_{50} with duration of exposure, for molluscs and crustaceans; EC_{50} values (concentration that is effective in 50% of test subjects), for microbes, algae and bacteria], but also fresh water and marine micro-organisms and plants, are reported. Persistence and potential for accumulation in the environment (bioaccumulation) and any available information on the harmful effects to non-target species, *i.e.* the unintentional exposure of terrestrial and/or aquatic species to a toxic substance, is given.

Environmental fate. The data in the environmental fate section deserves further expansion to explain the data that is quoted and the use of that data for an environmental assessment.

Nitrification inhibition is the inhibition of the nitrogen cycle, which is the major biogeochemical process in the production of nitrogen, an essential element contained in amino acids and proteins. The degree of inhibition can be used to estimate the environmental impact of the test chemical.

Carbonaceous inhibition is the recycling of carbon via the decomposition of complex organic matter by bacteria and fungi. In nature the process is important in the cycling of elements and nutrients in ecosystems. Chemical inhibition of microbial processes at all or any of these stages is given.

Anaerobic effects: Anaerobic microbial degradation of organic compounds occurs in the absence of oxygen and is an important degradation process in both the natural environment and waste treatment plants.

Degradation studies: Degradation data is used to assess the persistence of a chemical substance in the environment, in water, soil, and air. If the substance does not persist, information on the degradation products is also desirable. Intermediates may be either harmless or toxic substances, which will themselves persist. This section focuses on microbial degradation in both soil and water.

Abiotic removal reports the extent of photolytic and oxidative reactions occurring in the atmosphere and hydrolysis in water, which can be used as a measure of environmental pollution likely to arise from exposure to a substance.

Adsorption reports the extent to which a substance has an affinity to soil or sediment, water, or air and therefore the ability of the substance to move through the environment.

A sample entry from DOSE is given in Appendix C.

The Pesticide Manual (incorporating The Agrochemicals Handbook), 10th edn.[22] This latest edition of the Pesticide Manual, published in November 1994, was the result of an amalgamation of 'The Pesticide Manual' (British Crop Protection Council) and 'The Agrochemicals Handbook' (The Royal Society of

Chemistry). In the amalgamation the data has been extensively revised and updated to create a comprehensive reference volume geared to the diversity of interests in pesticides, including worldwide concern about the impact of pesticides on the environment.

'The Pesticide Manual' covers over 700 pesticide active ingredients registered in the USA and worldwide and over 500 superseded active ingredients, and chemical and biological control agents: herbicides, fungicides, plant growth regulators, and rodenticides.

The data on each active ingredient is arranged within the following groupings:

- Full nomenclature and structure
- Uses
- Product and residue analysis
- Physical and chemical properties
- Toxicology data
- Manufacturing companies
- Ecotoxicology
- Environmental fate data

The ecotoxicology data comprise:

- Toxicity to birds (LD_{50}/LC_{50}), fish (LC_{50}), bees (LD_{50}), *Daphnia* (EC_{50}), and other aquatic species.

The environmental fate data comprise:

- Animal data: biodegradation
- Plant data: biodegradation
- Soil and water: decomposition and persistence levels

A typical entry is reproduced in Appendix D.

References

1 'SIGLE (System for Information on Grey Literature in Europe)', The British Library, 1985–to date. Available on-line via Blaise-line and CD-ROM.
2 'Enviroline', Congressional Information Services (CIS), Bethesda, 1971–to date. Available on-line via Data-Star, KR DIALOG, DIMDI, ESA-IRS, Questel-Orbit, and on CD-ROM.
3 'Environmental Bibliography', International Academy at Santa Barbara, 1973–to date. Available on-line via KR DIALOG and on CD-ROM.
4 'Pollution Abstracts', Cambridge Scientific Abstracts (CSA), Bethesda, 1970–to date. Available on-line via Data-Star, KR DIALOG, ESA-IRS, STN International, and on CD-ROM.
5 'Aqualine', Water Research Centre (WRc) plc, Medmenham, 1960–to date. Available on-line via Questel-Orbit and on CD-ROM.

6 'Chemical Abstracts', Chemical Abstracts Service, Columbus, 1967–to date. Available on-line via STN International (with abstracts), and index on CD-ROM.

7 'Compendex (engineering database)', Engineering Information Inc. (EI), Hoboken, 1970–to date. Available on-line via Data-Star, KR DIALOG, DIMDI, ESA-IRS, STN International, and on CD-ROM.

8 'Cab Abstracts', CAB International (CABI), Wallingford, 1972–to date. Available on-line via Data-Star, KR DIALOG, DIMDI, ESA-IRS, STN International, and on CD-ROM.

9 'ECDIN (Environmental Chemicals Data Information Network)', Commission of the European Communities, Ispra, 1970–to date. Available via DIMDI and on CD-ROM.

10 'HSDB (Hazardous Substances Data Bank)', US National Library of Medicine (NLM), Bethesda, 1985–to date. Available via Data-Star, DIMDI, Medlars, and STN International.

11 'Oceanic Abstracts', Cambridge Scientific Abstracts (CSA), 1964–to date. Available on-line via Data-Star, ESA-IRS, and STN International.

12 'Toxline', US National Library of Medicine (NLM), 1965–to date. Available via Data-Star, KR DIALOG, Medlars, STN International, DIMDI, CCP online, and on CD-ROM.

13 R. E. Hester and R. M. Harrison, 'Issues in Environmental Science and Technology', The Royal Society of Chemistry, Cambridge.

14 'Chemical Safety NewsBase', The Royal Society of Chemistry, 1981–to date. Available on-line via KR DIALOG, Data-Star, STN International, and Questel-Orbit.

15 KR OnDisc 'Environmental Chemistry, Health & Safety', The Royal Society of Chemistry, 1981–to date. Available via KR OnDisc.

16 M. L. Richardson, 'Risk Management of Chemicals', The Royal Society of Chemistry, Cambridge, 1992.

17 A. J. Beck, K. C. Jones, M. H. B. Hayes, and U. Mingelgrin, 'Organic Substances in Soil and Water', The Royal Society of Chemistry, Cambridge, 1993.

18 J. A. G. Drake, 'Integrated Pollution Control', The Royal Society of Chemistry, Cambridge, 1994.

19 J. Paul, 'Carbon Dioxide Chemistry: Environmental Issues', The Royal Society of Chemistry, Cambridge, 1994.

20 J. A. G. Drake, 'The Chemical Industry – Friend to the Environment?', The Royal Society of Chemistry, Cambridge, 1992.

21 M. L. Richardson and S. Gangolli, 'The Dictionary of Substances and their Effects', The Royal Society of Chemistry, Cambridge, 1992–1995.

22 C. Tomlin, 'The Pesticide Manual incorporating The Agrochemicals Handbook', 10th edn, The British Crop Protection Council and The Royal Society of Chemistry, Cambridge, 1994.

Appendix A

Dictionaries

Earthwords: Dictionary of the Environment
Seymour Simon
Harper Collins Publishers, USA, 1995
ISBN 0-06020-234-3

Dictionary of Ecology and Environment
P H Collin
P Collin Pub. 1995
ISBN 0-94854-974-2

Dictionary of Environment and Development,
People, Places, Ideas and Organizations
Andy Crump
Earthscan, London, 1991
ISBN 1-85383-078-X

The Dictionary of Global Climatic Change
Compiled by W John Maunder
UCL Press Ltd, London, 1992
ISBN 1-85728-023-7

The Environmental Dictionary
Compiled by James J King
Executive Enterprises Publications Co. Inc.
New York NY, 2nd ed. 1993
ISBN 0-78460-171-1

Directories

ECO Directory of Environmental Databases in the United Kingdom
Monica Barlow and John Button
ECO Env. Info. Trust, Bristol, 1995
ISBN 1-8746-601-6

Directory of Environment and Development Issues
Network of Irish Environment & Development
Organisations,
c/o 10 Upper Camden Street,
Dublin 2, Republic of Ireland, 1995
ISBN 0-95255-210-8

Environmental Profiles. A global guide to projects and people
Linda Sobel, Sarah Orrick and Robert Honig
Garland Publishing Inc., New York and London, 1993
ISBN 0-81530-063-8

PIRA'S International Environmental Information Sources
Compiled by Susan Farrell in collaboration with Pira Information Centre
Paper, Packaging, Printing and Publishing Industry Research Association, 1990
ISBN 0-90279-958-4

Who's Who in the Environment, England
Edited and updated by Kate Aldous and Rachel Adatia with Kim Milton
The Environment Council, London, 3rd ed. 1995
ISBN 0-90315-852-3

IAWQ Yearbook 1995-96
Editor Heather Riley
International Association on Water Quality,
London, 1995
ISSN 1357-1729

Who's Who in European Water
Edited by Roy Harris
Sterling Publications Ltd, London, 1993
ISSN 0966-7083

Polmark, The European Pollution Control and Waste Management Industry Directory
Frost and Sullivan, London, 1991
ISBN 0-86354-589-0

The Environment Industry Yearbook 1995
The Environment Press Ltd, Bath, 3rd ed. 1994
ISBN 0-95190-960-3

The European Environmental Statistics Handbook
Oksana Newman and Allen Foster
Gale Research International Ltd, London, 1993
ISBN 1-87347-766-5

Statistical Record of the Environment
Compiled and edited by Arsen J Darnay
Gale Research Inc., London and Detroit, 2nd ed. 1994
ISBN 0-81038-863-4

Periodicals

The Ends Report
Environmental Data Services Ltd,
40 Bowling Green Lane, London EC1 0NE
ISSN 0966-4076

Environment Today
Brodie Publishing Ltd,
11–13 Victoria Street, Liverpool L2 5QQ
ISSN 0966-4130

Industry and Environment
United Nations Environment Programme
Industry and Environment Programme
Activity Centre, 39–43 Quai Andre' Citroen,
75739 Paris, Cedex 15, France
ISSN 0378-9993

Journal of Cleaner Production
Butterworth-Heinemann Ltd,
88 Kingsway, London WC2B 6AB
ISSN 0959-6526

Environmental Protection Bulletin
The Institution of Chemical Engineers
165–171 Railway Terrace, Rugby CV21 3HQ
ISSN 0957-9052

Pollution Equipment News
Rimbach Publishing Inc.
8650 Babcock Boulevard, Pittsburgh,
PA 15237-5821

Waste and Environment Today (news journal)
Editor Peter Doyle
AEA Technology, B7, 12 Harwell, Didcot,
Oxfordshire OX11 0RA

Industrial Waste Management
International Env. Technology and Pollution
Control Magazine, Faversham House Group Ltd.,
111 St James's Road, Croydon CR9 2TH
ISSN 0961-4719

Warmer Bulletin
World Action For Recycling Materials and Energy from Rubbish
Bridge House, High Street, Tonbridge TN9 1DP

Water Bulletin
Editor Matt Haddon
Water Services Association
1 Queen Anne's Gate, London SW1H 9BT
Registered as a newspaper

Marine Pollution Bulletin
Elsevier Science Ltd,
The Boulevard, Langford Lane,
Kidlington, Oxford OX5 1GB
ISSN 0025-326X

Environmental Education
National Association for Environmental Education
University of Wolverhampton, Walsall
Campus, Gorway, Walsall WS1 3BD
ISSN 0309-8451

Habitat
The Environment Council
21 Elizabeth Street, London SW1W 9RP
ISSN 0028-9043

Earth Matters
Friends of the Earth
26–28 Underwood Street, London N1 7JQ
ISSN 0956-6651

European Environment
European Research Press Ltd.
34–38 Chapel Street, Little Germany, Bradford BD1 5DN
ISSN 0961-0405

Environmental Auditing

Environmental Compliance Auditing
B Cleaver
S Thornes, 1995
ISBN 0-74872-187-8

Environmental Accounting and Auditing
John Collier
Prentice-Hall, 1995
ISBN 0-1335-5645-X

Auditing for Environmental Quality
Leadership: Beyond Compliance to Environmental Excellence
Wiley, 1995
ISBN 0-47111-492-8

Environmental, Health and Safety Auditing Handbook
Lee Harrison
McGraw, 1995
ISBN 0-07026-921-1

Environmental Accounting and Auditing
John Collier
Prentice-Hall, 1995
ISBN 0-13355-645-X

Guide to Local Environmental Auditing
Hugh Barton
Earthscan Publications, 1995
ISBN 1-85383-234-0

Environmental Auditing: An Introduction and Practical Guide
Editor Helen Woolston
The British Library, London, 1993
ISBN 0-71230-789-3

Environmental Auditing. A guide to best practice in the UK and Europe
Compiled by Lesley Grayson
Technical Communication and The British Library, London, 1992
ISBN 0-94665-558-8

BS7750: What the new environmental standards mean for your business
Compiled by Lesley Grayson
Technical Communication and The British Library, London, 1992
ISBN 0-94665-560-X

Waste Management. The duty of care, a code of practice
DOE, Scottish Office and Welsh Office
Environmental Protection Act 1990
HMSO, London, 1991
ISBN 0-11752-557-X

Environmental Auditor
Springer International
PO Box 580 East Lyme, Connecticut, USA
ISSN 0933-0437

Environmental Assessment
Institute of Environmental Assessment Magazine
R & W Publications Ltd., Goodwin House,
Willies Snaith Road, Newmarket CB8 7SQ
ISSN 1351-0738

Environmental Audit
Health and Safety Technology and Management
Mercury Books Ltd, London, 1991
ISBN 1-85252-100-7

Green Business
Malcolm Wheatley
Pitman Publishing, London, 1993
ISBN 0-27360-020-6

Narrowing the Gap: Environmental auditing guidelines for business
CBI, London, 1990
ISBN 0-85201-371-X

Business

Managing Environmental Crisis
Peter Bulleid
S Thornes, 1995
ISBN 0-74872-439-7

Company Environmental Reporting: A Measure of the Progress of Business and Industry
Towards Sustainable Development
United Nations Environment Programme
HMSO, London, 1995
ISBN 9-28071-413-9

Economic Development and Environmental Control Balancing Business and Community
in an Age of NIMBYs and LULUs
John O'Looney
Quorum Books, USA, 1995
ISBN 0-89930-940-2

Business and the Environment
Michael Flood
Powerful Information, 21 Church Lane,
Loughton, Milton Keynes MK5 8AS, 1995
ISBN 1-89995-001-X

Environmental Improvement Through the Management of Waste
D G Jones
S Thornes, 1995
ISBN 0-74872-129-0

Measuring Environmental Performance: A Comprehensive Guide
Walter Wehrmeyer
S Thornes, 1995
ISBN 0-74872-063-4

Environment Business
Editor Ian Grant
Information For Industry Ltd,
521 Old York Road, London SW18 1TG
ISSN 0959-7042

Environmental Management Systems
Institution of Chemical Engineers, UK, 1995
ISBN 0-85295-363-1

Greenpeace Business
Greenpeace Ltd.
John Sauven
Canonbury Villas, London N1 2PN
ISSN 0962-9467

International Environmental Policy and Management: Business and Emerging Markets
Linda S Spedding
S Thornes, 1995
ISBN 0-74872-132-0

Strategy for Sustainable Business; Environmental Opportunity and Strategic Choice
Liz Crosbie
McGraw, UK, 1995
ISBN 0-07709-133-7

Risk Assessment and Management Handbook for Environmental, Health and Safety
Professionals
Rao V Kolluru
McGraw, UK, 1995
ISBN 0-07035-987-3

Power of Environmental Partnerships
Frederick Long
HB Dryden, USA, 1995
ISBN 0-03011-327-X

Environmental Management in a Transition to Market Economy: A Challenge to
Governments and Business
Editions Technip, 27 Rue Ginoux, 75737 Paris
CEDEX 15, France, 1995
ISBN 2-71080-680-0

Environmental Marketing Management
Ken Peattie
Pitman, UK, 1995
ISBN 0-27360-279-9

Environmental Management Systems:
Principles and Practice
David Hunt
McGraw, 1995
ISBN 0-07707-910-8

Improving Environmental Performance
Suzanne Pollack
Routledge, 1995
ISBN 0-41510-237-5

Narrowing the Gap: Environmental auditing guidelines for business
CBI, London, 1990
ISBN 0-85201-371-X

Environmental Policy Benefits: Monetary Valuations
Prepared by D Pearce and A Markandya.
Contribution from J-P H Barde
Organisation for Economic Co-operation and Development, Paris, 1989, 83p.
ISBN 9-26413-182-5

World Guide to Environmental Issues and Organisations
Edited by Peter Bracklet
Longman, Harlow, Essex, 1990
ISBN 0-58206-270-5

Corporate Environmental Register
The Environment Press, London, 1993
ISBN 0-95190-961-4

Environmental Pollution Risks. A Practical Guide
Edited by G Village
DYP Group, London, 1993

Environmental Decision Making: A Multidisciplinary Perspective
Edited by R Chechil and S Carlisle
Van Nostrand Reinhold, New York, 1991, 296p
ISBN 0-44200-659-4

The Green Management Revolution: Lessons in environmental excellence
W Hopfenbeck
Prentice Hall, Herts., 1993, 326pp.
ISBN 0-13276-452-0

The Science of Global Change: The impact of human activities on the environment
Edited by D A Dunnette and R J O'Brien
American Chemical Society, Washington, DC
ISBN 0-84122-197-9

Pollution

Volatile Organic Compounds in the Atmosphere
Edited by R E Hester and R M Harrison
Royal Society of Chemistry, UK, 1995
ISBN 0-85404-215-6

Computer Treatment of Large Air Pollution Models
Zahari Zlatev
Kluwer Academic, Dordrecht 1995
ISBN 0-79233-328-4

Waste Treatment and Disposal
Edited by R E Hester and R M Harrison
Royal Society of Chemistry, UK, 1995
ISBN 0-85404-210-5

Water Sampling for Pollution Regulation
Keith D Harsham
Gordon & Breach and Wepf & Co., Switzerland, 1995
ISBN 2-88449-040-X

Air Pollution
P Zannetti, C A Brebbia, J E Garcia Gardia and G Ayala Milian
Elsevier Science Publishers, London, 1993, 793pp.
ISBN 1-85312-222-X

Pollution Prevention
L Theodore and Y C McGuinn
Chapman Hall, London, 1992, 366pp.
ISBN 0-44200-606-3

Industrial Pollution Prevention Handbook
Edited by Harry M Freeman
McGraw-Hill, New York, London 1995
ISBN 0-07022-148-0

Ground Pollution, Environment, Geology, Engineering and Law
P Attewell
Chapman Hall, London, 1993, 251pp.
ISBN 0-41918-320-5

Limiting Greenhouse Effects: controlling carbon dioxide emissions
Editor G I Pearman
John Wiley, Chichester, 1992
ISBN 0-47192-945-X

Assessment of Noise Impact on the Urban Environment: A study on noise production models
J Lang
WHO, Regional Office for Europe,
Copenhagen, 1986, 71pp.
Environmental Health Series 9

Technology

Cleaner Technology
Her Majesty's Stationery Office, London,
1992, 72pp.
ISBN 0-11430-069-0

Green Gold: Japan, Germany, the US and the Race for Environmental Technology
Curtis Moore
Beacon Publishing, USA, 1995
ISBN 0-80708-531-6

Environmental Hydrology
Kluwer Academic, Dordrecht, 1995
ISBN 0-79233-549-X

Sustainable Development: Science, Ethics and Public Policy
John Lemons
Kluwer Academic, Dordrecht, 1995
ISBN 0-79233-500-7

Technology and the Future of Europe: Global competition and the environment in the 1990s
C Freeman, M Sharp and W Walker
Pinter, London, 1991

Guidelines for Assessing Industrial Environmental Impact and Environmental Criteria for the Siting of Industry
Industry and Environment Programme
UN Environment Programme, Moscow, 1985
ISBN 9-28071-015-X

Dust and Fume Control: A user guide
Editor D M Muir
Institution of Chemical Engineers, Rugby, 2nd rev. ed. 1992, 162pp.
ISBN 0-85295-287-2

Green Design
P Bunell
The Design Council, London, 1991, 81pp.
ISBN 0-85072-284-5

Green Design. Design for the Environment.
D Mackenzie
Lawrence King Ltd, London, 1991
ISBN 1-85669-001-6

Energy Efficiency for Engineers and Technologists
T D Eastop and D R Craft
Longman Scientific and Technical, Harlow, 1990
ISBN 0-58203-184-2

Recycling and Waste: An exploration of contemporary environmental policy
Matthew Gandy
Avenbury, Aldershot, 1993

Law

Direct Effect of European Law and the Regulation of Dangerous Substances
Christopher J M Smith
Gordon & Breach, Switzerland, 1995
ISBN 2-88449-042-6

Environmental Policy: From Regulation to Economic Instruments
Centre d'Etude et de Recherche de Droit
International et de Relations Internationales de l'Academie de Droit
Nijhoff. (In French and English) 1995
ISBN 0-79233-555-4

Trading Up: Consumer and Environmental
Regulation in a Global Economy
David Vogel
Harvard UP, USA, 1995
ISBN 0-67490-083-9

Economic Analysis of Environmental Policy and Regulation
Frank S. Arnold
Wiley, 1995
ISBN 0-47100-084-1

Chemistry and Environmental Legislation,
Methodologies and Applications
Edited by S Facchetti and D Pitea
Kluwer Academic, Dordrecht, 1995
ISBN 0-79233-240-7

The 1994 Waste Management Licensing
Regulations and the new Definition of Waste
Edited by Pamela Castle
Stanley Thornes, Cheltenham, 1995
ISBN 0-74872-133-9

Non Point Source Pollution Regulation Issues and Analysis
Edited by Cesare Dosi and Theodore Tomasi
Kluwer Academic, Dordrecht, 1994
ISBN 0-79233-121-4

Ground Pollution, Environment, Geology, Engineering and Law
P Attewell
Chapman & Hall, London, 1993, 251pp.
ISBN 0-41918-320-5

Manual of Environmental Policy: The EC and Britain
Edited by N. Haigh
Longman, London, 1991, Loose-leaf
ISBN 0-58208-715-5

Garners Environmental Law
Butterworths, Loose-leaf, in 3 vols.
ISBN 0-40620-560-4

European Community Environment Legislation
Commission of the European Communities
Directorate General X1
Luxembourg, 1992
ISBN 9-28264-084-1

European Environmental Law Review
Graham and Trotman, London
ISSN 0966-1646

Journal of Environmental Law
Oxford University Press, Eynsham
ISSN 0952-8873

Clean Air Round the World
Editor L Murley
International Union of Air Pollution
(IUAPPA), Brighton, 2nd ed. 1991
ISBN 1-87168-801-9

Environmental Protection Act 1990: Waste management, the duty of care, a code of
practice.
DOE, Scottish Office and Welsh Office.
Her Majesty's Stationery Office, London, 1991
ISBN 0-11752-557-X

National Environmental Waste Policy Conference (House of Lords April 23, 1990) and
Waste Policy Review (Report) Paper, Packaging
Printing and Publishing Industry Research Association (PIRA) in Collaboration with
Strategy Europe LTS, Leatherhead, 1990
ISBN 0-90279-931-2

Environmental Protection Act 1990
Narrowing the Gap: Environmental auditing guidelines for business
Confederation of British Industry, London, 1990
ISBN 0-85201-371-X

Planning and Environmental Law Bulletin
Longman, Harlow
ISSN 0962-4597

Energy

Environment and Energy
Edited by T Nejat Veziroglu
Nova Science Publishers, New York, 1991
ISBN 1-56072-001-8

Energy and Environment in the European Union: The challenge of integration
Ute Collier
Avebury Studies in Green Research,
Aldershot, 1994
ISBN 1-85972-007-2

Energy, Physics and the Environment
Ernest L McFarland, James L Hunt, John L Campbell
Wuerz, Winnipeg, 1994
ISBN 0-92006-362-4

Biofuels
International Energy Agency, Paris, 1994
ISBN 9-26414-233-9

Energy Policy in the Greenhouse
Florentin Krause
Earthscan Publications Ltd, London, 1989, 1990

Energy and the Environment
Editor Bryan Cartledge
Oxford University Press, Oxford, 1993
ISBN 0-19858-419-9 (pbk)

Power Generation and the Environment
L E J Roberts, P S Liss and P A H Saunders
Oxford University Press, Oxford, 1990
ISBN 0-19858-338-9

Appendix B: Environmental Organizations

from Portico – The British Library Online Information Service

http://portico.bl.uk/sris/eis/orgs1.html

Alfred-Wegener-Institute   **http://www.awi-bremerhaven.de/**

The Alfred-Wegener-Institute is a German national research centre for Polar and Marine research.

Alpha Analytical Labs   **http://world.std.com/~alphalab/**

Alpha is a full service environmental analytical laboratory.

AquaNIC   **http://thorplus.lib.purdue.edu/Aquanic/ahome2.html**

The Aquaculture Network Information Center (AquaNIC) is intended to be a gateway to the world's electronic resources in aquaculture.

Australian Oceanographic Data Centre   **http://www.Aodc.Gov.Au/aodc.html**

The primary activity of the AODC is the development of Navy's Marine Environmental Database (MEDB) which provides the basis of most products produced within AODC. The AODC is also responsible for acquiring, managing and disseminating marine environmental data to the civilian marine science community and the general public.

Base de Dados Tropical from Brazil   **http://www.ftpt.br/**

The Base de Dados Tropical (Tropical Data Base – BDT) is a department within the Fundacao Tropical de Pesquisas e Tecnologia "Andre' Tosello", a Brazilian not-for-profit, private foundation.

BENE   **http://straylight.tamu.edu/bene/bene.html**

The Biodiversity and Ecosystems Network, BENE, is designed to foster enhanced communications and collaborations among those interested in biodiversity conservation and ecosystem protection, restoration, and management.

British Columbia – Environment   **http://www.env.gov.bc.ca/**

British Columbia Ministry of Environment, Lands and Parks World Wide Web Server.

CEDAR: Central European Environmental Data Request Facility

http://pan.cedar.univie.ac.at/

Provides computing and Internetwork facilities to support international data exchange with the Central and Eastern European environmental community.

CERES: California Environmental Research Evaluation System

http://agency.resource.ca.gov/

CERES is an information system being developed by the Resources Agency to facilitate access to a variety of electronic data describing California's rich and diverse resources.

CIESIN **gopher://infoserver.ciesin.org/**

The Consortium for International Earth Science Information Network (CIESIN, pronounced "season") was established in 1989 as a private, nonprofit membership corporation with members from leading universities and non-government research organizations.

Cornell University Center for the Environment **http://www.cfe.cornell.edu/**

The Center for the Environment (CfE) addresses pressing environmental issues in their full interdisciplinary complexity, through teaching, research and outreach.

The Daily Planet **htp://www.atmos.uiuc.edu/**

Brought to you by the Department of Atmospheric Sciences at the University of Illinois. This is the location of our fully developed environmental information server (EIS), The Daily Planet. In addition to weather images and MPEG animations, there are links to our online electronic textbook, and a number of other locally developed resources.

Earthwatch **http://gaia.earthwatch.org**

Earthwatch sponsors expeditions that improve the quality and management of life's resources.

ECN **http://www.nmw.ac.uk/ecn/**

Founded in 1992, ECN is the UK's integrated long-term environmental monitoring network. It is designed to collect, store, analyse and interpret long-term data based on a set of key physical, chemical and biological variables which drive and respond to environmental change.

EcoNet **http://www/igc.apc.org/igc/en.html**

EcoNet serves organizations and individuals who are working for environmental preservation and sustainability.

EcoWeb **http://ecosys.drdr.virginia.edu/EcoWeb.html**

EcoWeb is devoted to facilitating access to local recycling and environmental information as well as more comprehensive environmental resources.

EKN Canada **http://ekn.sid.ncr.doe.ca/**

Environment Canada is committed to supporting Canadians in their efforts to sustain the environment to ensure that a positive environmental legacy is passed on to future generations.

Energy & Environmental Research Center   **http://eerc.und.nodak.edu/**

The EERC is one of the world's leading energy and environmental facilities.

Environment Canada Green Lane   **http://www.ns.doe.ca/**

The Green Lane will provide interactive access to Environment Canada services, products, information holdings, programs and policies.

Environment at MIT   **http://web.mit.edu/org/c/ctpid/www/tbe/tbe-home.html**

The Technology, Business and Environment Program was founded to help companies meet the dual challenges of achieving environmental excellence and business success.

Environmental Web Resources   **http://envirolink.org/envirowebs.html**

The Environmental World Wide Web Listing at EnviroLink. This is a listing of all of the environmental World Wide Web services that we are aware of.

EPA WWW   **http://www.hpcc.gov/blue94/section.4.9.html**

Use EPA web server.

Environmental Resource Centre   **http://ftp.clearlake.ibm.com/erc/homepage.html**

The Environmental Resource Center (ERC) is an innovative cooperative between private industry and multiple levels of government. Directed and developed by industry, the ERC will provide an effective means to assimilate, enhance, and distribute existing environmental knowledge.

EPA Environmental Gopher   **gopher://gopher.epa.gov/11/other/gophers**

US Environmental Protection Agency Gopher.

EnviroWeb   **http://envirolink.org/**

The EnviroWeb is the EnviroLink Network's World Wide Web server.

ERIN Australia   **http://kaos.erin.gov.au/erin.html**

The Environmental Resources Information Network [ERIN] aims to provide geographically related data of an extent, quality and availability required for planning and decision making.

FireNet Information Network   **http://www.anu.edu.au/forestry/fire/firenet.html**

FireNet as an on-line information service for everyone interested in rural and landscape fires. The information concerns all aspects of fire science and management – including fire behaviour, fire weather, fire prevention, mitigation and suppression, plant and animal responses to fire and all aspects of fire effects.

Friends of the Earth   **http://www.foe.co.uk/**

The largest international network of environmental groups in the world, represented in 52 countries and one of the leading environmental pressure groups in the UK.

Gateway to Antarctica   **http://icair.iac.org.nz/**

The International Centre for Antarctic Information and Research (ICAIR) collects, analyses, distributes and co-ordinates scientific, environmental and educational information relating to Antarctica. ICAIR is an international, politically neutral institution incorporated within the science academy The Royal Society of New Zealand.

GENIE   **http://www-genie.mrrl.lut.ac.uk/**

The GENIE project will provide a user-sympathetic system for locating and accessing relevant information on Global Environmental Change.

Global Futures Foundation   **http://www.quicknet.com/globalff/globalfu.html**

Global Futures Foundation (GFF), an innovative environmental non-profit foundation. GFF focuses on systematically integrating programs which lead to source reduction, pollution prevention, low-cost market development, and incentive market driven regulatory structures which tend to reduce both economic and environmental costs.

Global Recycling Network   **http://grn.com/grn**

GRN is the most comprehensive Recycling Information Resource available on the Internet. Besides a wide variety of general recycling reference material, GRN offers a virtual market-place intended to help businesses around the world in finding possible trading partners for the sale of recyclable goods.

Greenpeace International (Amsterdam)   **http://www.greenpeace.org/**

Greenpeace International is the international coordinating body for the 43 national offices in 30 countries.

ICE: University of California   **http://ice.ucdavis.edu/#top of list**

The Information Center for the Environment (ICE) is a cooperative effort of an interdepartmental team of environmental scientists at the University of California, Davis and collaborators at over thirty private, state, federal, and international environmental organizations.

INFOTERRA   **http://pan.cedar.univie.ac.at/gopher/unep/unep.html**

INFOTERRA The Global Environmental Information Exchange Network, established in 1975 by a decision of the third session of the Governing Council of UNEP. The main direction given to INFOTERRA was to develop a mechanism to "facilitate the exchange of environmental information within and among nations".

Institute of Terrestrial Ecology   **http://www.nmw.ac.uk:80/ite/**

One of four Institutes which form the Centre of Ecology and Hydrology, a part of the UK Natural Environment Research Council responsible for research into all aspects of the terrestrial environment and its resources.

LEAD International Leadership for Environment and Development Program

http://www.lead.org/

Linkages   **http://www/mbnet.mb.ca:80/linkages/**

Linkages is provided by the International Institute for Sustainable Development (IISD), publishers of the Earth Negotiations Bulletin. It is designed to be an electronic clearing-house for information on past and upcoming international meetings related to environment and development.

National Environmental Information Resources Center

http://www.gwu.edu/~greenu/

National Institute for the Environment   **http://www.inhs.uiuc.edu/cnie.html**

The Committee for the National Institute for the Environment (CNIE) is a national, non-profit organization working to improve the scientific basis for making decisions on environmental issues, through creation of a new, non-regulatory environmental science and education agency, the National Institute for the Environment (NIE).

Natural Environment Research Council   **http://www.nerc.ac.uk/**

The mission of the Natural Environment Research Council is: to promote and support, by any means, high quality basic, strategic and applied research, survey, long-term environmental monitoring and related post-graduate training in terrestrial, marine and freshwater biology and Earth, atmospheric, hydrological, oceanographic and polar sciences and Earth observation;

Natural History Museum   **http://www.nhm.ac.uk**

The Natural History Museum is dedicated to furthering the understanding of the natural world through its unrivalled collections, its world class exhibitions and education, and through its internationally significant programme of scientific research.

NOAA ESDIM Home Page   **http://www.esdim.noaa/gov/**

National Oceanic and Atmospheric Administration Environmental Information Services.

NREL's Gopher/WWW Server   **http://www.nrel.gov/**

National Renewable Energy Laboratory, a national laboratory of the US Department of Energy.

Oak Ridge National Laboratory (ORNL)   **http://www.ornl.gov/**

ORNL is a Department of Energy multiprogram laboratory managed by Lockheed Martin Energy Systems, Inc. Scientists at ORNL to conduct a wide range of basic and applied research and development to advance in several Core Competencies the nation's energy resources, environmental quality, scientific knowledge, educational foundations, and economic competitiveness.

Oceanic Information Centre   telnet://delocn.udel.edu/info

Pesticide Action Network North America (PANNA)
                                              gopher://gopher.igc.apc.org/11/orgs/panna

The PANNA Update Service (PANUPS) is a weekly news service featuring articles on pesticide use and sustainable agriculture from around the world, as well as action alerts and conference reports. The Pesticide Information Service (PESTIS) is an online database that contains pesticide reform-related material generated by NGOs, including articles, newsletters, reports and action alerts, all of which can be full-text searched.

Royal Botanic Gardens/Kew   http://www.rbgkew.org.uk/

The mission of the Royal Botanic Gardens, Kew is to enable better management of the earth's environment by increasing knowledge and understanding of the plant kingdom.

Seashepherd Gopher   gopher://envirolink.org/11/enviroorgs/.eorgs/.seashepherd

The Sea Shepherd Conservation Society is a non-profit organization involved with the investigation and documentation of violations of international laws, regulations and treaties protecting marine wildlife species.

Solstice   http://solastice.crest.org

Solstice, the site for energy efficiency, renewable energy, and sustainable technology information and connections.

Sustainable Earth Electronic Library   http://www.envirolink.org/pubs/

A unique collection devoted exclusively to materials that educate people on ways to preserve and restore our natural environment. The Sustainable Earth Electronic Library (SEEL) is a project of Sustainable Earth, Inc., a nonprofit organization devoted to the creation of environmentally-related information tools and services.

UK Government (CCTA)   http://www.open.gov.uk/

CCTA is the UK Government Centre for Information Systems, part of the Office of Public Service and Science, which works to improve government's services to the public. It contains information from:

- Ministry of Agriculture, Fisheries and Food (MAFF)
- Countryside Commission
- The Department of the Environment (DoE)
- Department of Trade and Industry (DTI)
- Her Majesty's Inspectorate of Pollution (HMIP)
- Ordnance Survey
- Pesticides Safety Directorate

United Nations   **gopher://nywork1.undp.org/**

United Nations Gopher Server.

United Nations Development Programme   **http://www.undp.org/**

United Nations Development Programme World Wide Web Server.

Universities Water Information Network   **http://www.unwin.siu.edu/**

UNWIN is designed to aid the flows of water information along the information superhighway. UNWIN maintains many information services of interest to managers, researchers, consultants, and teachers throughout the water resources community.

University of East Anglia: Environment   **http://www.env.uea.ac.uk**

In the School of Environmental Sciences physical, chemical, biological and social science methods are applied to the study of natural and human environments and man's role in them.

Virginia: Environment WWW Virtual Library Environment
                           http://ecosys.drdr.virginia.edu/environment.html

World Conservation Monitoring Centre   **http://www.wcmc.org.uk/**

WCMC provides information services on the conservation and sustainable use of species and ecosystems, and supports others in the development of their own information management systems.

World Health Organization   **http://www.who.ch/**

The objective of WHO is the attainment by all peoples of the highest possible level of health. Health, as defined in the WHO Constitution, is a state of complete physical, mental and social well-being and not merely the absence of disease or infirmity.

Appendix C: DOSE Item

p-Toluidine

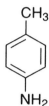

CAS Registry No. 106-49-0
Synonyms 4-methylbenzenamine; 4-aminotoluene; 4-methylaniline; C.I. 37107;
C.I. Azoic Coupling Component 107; *p*-tolylamine
Mol. Formula C_7H_9N **Mol. Wt.** 107.16
Uses Manufacture of dyes and other organic chemicals. Reagent for lignin, nitrite and
phloroglucinol.

Physical properties
M. Pt. 44-45°C; **B. Pt.** 20-201°C; **Flash point** 86°C; **Specific gravity** d_4^{20} 1.046;
Partition coefficient log P_{ow} 1.39; **Volatility** v.p. 1 mmHg at 42°C; v. den. 3.9.

Solubility
Water: soluble in 135 parts water. Organic solvent: ethanol, diethyl ether, acetone,
methanol

Occupational exposure
US TLV (TWA) 2 ppm (8.8 mg m^{-3}); **UN No.** 1708; **HAZCHEM Code** 3X;
Conveyance classification toxic substance; **Supply classification** toxic.
Risk phrases Irritating to eyes (R23/24/25, R33)
Safety phrases After contact with skin, wash immediately with plenty of soap and
water – Wear suitable protective clothing and gloves – If you feel unwell, seek
medical advice (show label where possible) (S28, S36/37, S44)

Ecotoxicity

Fish toxicity
Exposure to 5 ppm was non-toxic to bluegill sunfish, yellow perch and goldfish. Test
conditions: temperature 30°C; dissolved oxygen 7.5 ppm; total hardness (soap
method) 300 ppm; alkalinity 310 ppm (methyl orange); free carbon dioxide 5 ppm (1).

Invertebrate toxicity
EC$_{50}$ (30 min) *Photobacterium phosphoreum* 4.27 ppm Microtox test (2).

Environmental fate

Degradation studies

Confirmed to be biodegradable (3).

ThOD 2.54 g O_2 g^{-1}; BOD_5 1.44 g O_2 g^{-1} (4).

Decomposition by a soil microflora in 4 days (5).

Activated sludge at 20°C with compound as sole carbon source COD 97.7%; 20 mg COD g dry inocculum^{-1} hr^{-1} (6).

Degradation by *Aerobacter* 500 mg l^{-1} at 30°C; parent 100% ring disruption in 3 hr (7).

Mammalian and avian toxicity

Acute data

LD_{50} oral redwing blackbird, starling 56.2, 42.2 mg kg^{-1} respectively (8).

LD_{50} oral quail 237 mg kg^{-1} (8).

LD_{50} oral mouse, rat 330, 656 mg kg^{-1}, respectively (9,10).

LD_{50} intraperitoneal mouse 50 mg kg^{-1} (11).

Irritancy

Dermal rabbit (24 hr) 500 mg caused severe irritation, and 100 mg instilled in rabbit eye caused severe irritation (10).

Dermal rabbit (24 hr) 500 mg caused mild irritation, and 20 mg instilled in rabbit eye (24 hr) caused moderate irritation (12).

In rats exposure to 640 mg m^{-3} caused nasal and eye irritation (duration unspecified) (13).

Sensitisation

Caused sensitisation in guinea pigs (14).

Genotoxicity

Salmonella typhimurium TA98, TA100, TA1535, TA1537, TA1538, D3052, C3076, G46 with and without metabolic activation negative (15).

Escherichia coli WP2, WP2 *uvrA⁻* with and without metabolic activation negative (15).

In vitro primary rat hepatocytes unscheduled DNA synthesis positive (15).

In vitro V79 Chinese hamster lung cells did not cause single strand DNA breaks (16).

Any other comments

Industrial hazards reviewed (17).

Human health effects, experimental toxicology, physico-chemical properties, workplace experience, ecotoxicology, epidemiology reviewed (18).

Autoignition temperature 482°C.

References

1. *The Toxicity of 3400 Chemicals to Fish* 1987, EPA560/6-87-002 P13 87-200-275, Washington, DC
2. Kaiser, K. L. E. et al *Water Pollut. Res. J. Can.* 1991, **26**(3), 361-431
3. *The list of the existing chemical substances tested on biodegradability by microorganisms or bioaccumulation in fish body* 1987, Chemicals Inspection and Testing Institute, Japan
4. Meinck, F. et al *Les eaux residuaires indsutrielles* 1970
5. Alexander, M. et al *J. Agric. Food Chem.* 1966, **14**, 410
6. Pitter, P. *Water Res.* 1976, **10**, 231-235
7. Worne, H. E. *Tijdschrift van Het BECEWA* Liege, Belgium
8. Schafer, E. W. *Arch. Environ. Contam. Toxicol.* 1983, **12**, 355-382

9. *Gig. Tr. Prof. Zabol* 1981, **25**(8), 50
10. *BIOFAX Industrial Bio-Test Laboratorie Inc. Data Sheets* 1973, 31-4
11. *National Technical Information Service* AD691-490
12. Marhold, J. V. *Sbornik Vysledku Toxixologickeho Vysetreni Latek A Pripravku* 1972, Prague
13. *Chemical Safety Data Sheets* 1991, **4b**, 225-227, RSC, London
14. Kleniewska, D. et al *Dermatosen, Beruf, Umwelt* 1980, **28**, 11-13
15. Thompson, C. Z. et al *Environ. Mutagen.* 1983, **5**, 803-811
16. Zimmer, D. et al *Mutat. Res.* 1980, **77**, 371-326
17. Ikeda, M. et al *Sumitomo Sangyo Eisei* 1985, **21**, 131-151
18. *ECETOC Technical Report No. 30(5)* 1994, European Chemical Industry Ecology and Toxicology Centre, B-1160 Brussels

Appendix D: Pesticide Manual Item

dichlorvos   *Insecticide Acaricide*

organophosphorus

$$Cl_2C=CHO\overset{\overset{\displaystyle O}{\|}}{P}(OCH_3)_2$$

NOMENCLATURE
Common name dichlorvos (BSI, E-ISO, (*m*) F-ISO, BAN, ESA), DDVP (JMAF),
dichlorfos (USSR), *DDVF* (former exception USSR).
IUPAC name 2,2-dichlorovinyl dimethyl phosphate.
C.A. name 2,2-dichloroethenyl dimethyl phosphate.   **CAS RN** *[62-73-7]*
Development code Bayer 19 149; C 177   **Official code** OMS 14; ENT 20 738.

PHYSICO-CHEMICAL PROPERTIES
Mol. wt. 221.0   **Mol. formula** $C_4H_7Cl_2O_4P$
Form Colourless liquid; (tech., colourless-to-amber liquid with an aromatic odour).
B.p. 234.1 °C (OECD 102)   **V.p.** 2.1 Pa (25 °C) (OECD 104)   **SG/density** 1.425 (20 °C)
(OECD 109)   **K_{ow} logP** = 1.9 (OECD 117); 1.423 (separate study)   **Solubility** In water
c. 8 g/l (25 °C). Completely miscible with aromatic hydrocarbons, chlorinated
hydrocarbons and alcohols; moderately soluble in diesel oil, kerosene, isoparaffinic
hydrocarbons, and mineral oils.   **Stability** Stable to heat. Slowly hydrolysed in water and
in acidic media, and rapidly hydrolysed by alkalis, to dimethyl hydrogen phosphate and
dichloroacetaldehyde; DT_{50} (estimated) 31.9 d (pH 4), 2.9 d (pH 7), 2.0 d (pH 9) (22 °C).

COMMERCIALISATION
History Insecticide described by Ciba AG (now Ciba-Geigy AG) (GB 775085), but an
incorrect structure given to the compound; later reported by A. M. Martson *et al.* (*J.
Agric. Food Chem.*, 1955, **3**, 319) as an insecticidal impurity in trichlorfon. Introduced by
Ciba-Geigy AG, Shell Chemical Co. (now American Cyanamid Co.), and Bayer AG.
Patents GB 775085 to Ciba-Geigy; US 2956073 to Shell   **Manufacturer** Amvac; Bayer;
Bharat; Chemol; Ciba-Geigy; Cyanamid; Defensa; Denka; Jin Hung; Makhteshim-Agan;
Montecinca; Nippon Soda; Q.E.A.C.A.; Rhône-Poulenc; Sanachem.

APPLICATIONS
Mode of action Insecticide and acaricide with respiratory, contact, and stomach action.
Gives rapid knockdown. Cholinesterase inhibitor.   **Uses** Control of household and public

health insect pests, e.g. flies, mosquitoes, cockroaches, bedbugs, ants, etc.; stored-product pests in warehouses, storerooms, etc.; flies and midges in animal houses; sciarid and phorid flies in mushrooms; sucking and chewing insects, and spider mites in a wide range of crops, including fruit, vines, vegetables, ornamentals, tea, rice, cotton, hops, glasshouse crops, etc. Also used as a veterinary anthelmintic. **Phytotoxicity** Non-phytotoxic when used as directed, except to some varieties of chrysanthemum. **Formulation type** EC; AE; GR; HN; KN; Impregnated strip; OL. **Compatibility** Compatible with many other pesticides, but incompatible with alkaline materials, chinomethionat, and dichlofluanid. **Principal tradename** 'Dedevap' (Bayer), 'Nogos' (Ciba-Geigy), 'Vapona' (agronomic uses only) (Cyanamid), 'Didivane' (Diachem), 'Divipan' (Makhteshim-Agan), 'Phosvit' (Nippon Soda), 'Swing' (Siapa), 'Uniphos' (Chemol). **Mixtures** [*dichlorvos* +] propoxur; hexythiazox; isoxathion; phosalone; iodofenphos; propetamphos; piperonyl butoxide + pyrethrins; cyfluthrin + propoxur; methoprene + propoxur.

ANALYSIS
Product analysis by i.r. spectrometry (*AOAC Methods,* 1990, 964.04 966.07), by reaction with excess of iodine which is estimated by titration (*CIPAC Handbook,* 1980, **1A**, 1214) or by glc (*CIPAC Proc.,* 1981, **3**, 173). **Residues** determined by glc (*Pestic. Anal. Man.,* 1979, **I**, 201-H, 201-I; *Analyst* [*London*], 1973, **98**, 19; 1977, **102**, 858; 1980, **105**, 515; *Man. Pestic. Residue Anal.,* 1987, **I**, 3, 6, S13, S17, S19; *Anal. Methods Residues Pestic.,* 1988, Part I, M2, M5). Sampling of atmospheres (*Anal. Methods Pestic. Plant Growth Regul.,* 1972, **6**, 529). Methods for the determination of residues are available from Bayer.

MAMMALIAN TOXICOLOGY
Reviews *Food Cosmet. Toxicol.* 1974, **28**, 765-772 and A. S. Wright *et al., Arch. Toxikol.* 1979, **42**, 1-18. *Environmental Health Criteria* 79 (WHO, 1989). **IARC** 20, 53. **Acute oral** LD$_{50}$ for rats *c.* 50 mg/kg. **Skin and eye** Acute percutaneous LD$_{50}$ for rats *c.* 300 mg/kg. Skin and eye irritant (rabbits). **Inhalation** LC$_{50}$ (4 h) for rats > 0.1 mg/l air (vapour), *c.* 0.5 mg/l air (aerosol). **NOEL** (2 y) for rats 10 mg/kg diet. **ADI** (JMPR) 0.004 mg/kg b.w. [1993]. **Toxicity class** WHO Ib; EPA I.

ECOTOXICOLOGY
Birds Acute oral LD$_{50}$ for Japanese quail 26.8 mg/kg. **Fish** LC$_{50}$ (96 h) for rainbow trout 930, golden orfe 450 µg/l (both 500 EC). **Bees** Toxic to bees. **Daphnia** EC$_{50}$ (48 h) 0.19 µg/l. **Other aquatic spp.** EC$_{50}$ (5 d) for *Scenedesmus subupicatus* 52.8 mg/l.

ENVIRONMENTAL FATE
Animals In mammals, following oral administration, rapidly degraded in the liver by hydrolysis and *O*-demethylation, with a half-life of *c.* 25 minutes (L. Bull and R. L. Ridgway *J. Agric. Food Chem.* 1969, **17**, 837; D. H. Hutson and E. C. Hoadley *Arch. Toxikol.* 1972, **30**, 9-18; A. S. Wright *et al., Arch. Toxikol.* 1979, **42**, 1-18). **Plants** Rapidly decomposed in plants (D. L. Bull and R. L. Ridgway *J. Agric. Food Chem.* 1969, **17**, 837). **Soil and water** Non-persistent in the environment, with rapid decomposition in the atmosphere. Undergoes hydrolysis in damp media, with the formation of phosphoric acid and CO_2. Half-lives < 1 d in biologically active soils and water systems.

Chapter 13

The Acquisition of Environmental Data for Legislative Purposes

By John L. Vosser

FORMERLY, DEPARTMENT OF THE ENVIRONMENT,
ROMNEY HOUSE, 43 MARSHAM STREET, LONDON SW1P 3PY, UK

1   Introduction

This chapter gives the background to the legislative requirements of the European Union for the testing of chemicals for their environmental effects. Application of the test results is discussed. The requirements for the environmental testing of plant protection substances and non-agricultural pesticides are not covered but they have certain common elements with those for industrial chemicals.

2   Background to the European Union Requirements

A basic premise of the European Union (EU) is that technical and administrative barriers to trade should be eliminated. This applies to chemicals, and to achieve this, a harmonized system of testing, classification, labelling, and assessment of risk is needed. In practical terms this avoids needless duplication of effort since what is acceptable in one Member State will be acceptable to all of the others.

The testing of chemicals within the EU has its origins in a Directive adopted in 1967, 67/548/EEC,[1] which harmonized the criteria for the classification, packaging, and labelling (CPL) of certain dangerous chemicals. Neither the classification *etc.* for the environment nor any specific testing procedures were included.

3   New Chemicals

The first five amendments of this 1967 Directive were concerned with CPL, but the sixth one, Directive 79/831/EEC,[2] introduced a pre-marketing notification scheme for certain new chemicals. It continued the CPL requirements for old substances as well as those newly notified. Whilst the classification 'Dangerous for the Environment' was included in the list of categories of danger in this sixth amendment, environmental criteria had not been developed at the time it was adopted. The testing requirements included those for effects on certain species and fate in the environment.

A 'new chemical' for the purposes of this sixth amendment directive was defined as one which had not been commercially exploited in the European Community for the 10 years prior to adoption in September 1979. With the aid of the chemical industry, the European Commission compiled a list of chemicals and other substances which had been on the European Community market for the previous 10 years. Over 100 000 entries were made in this European Inventory of Commercial Chemical Substances, (EINECS). This list is now closed and intending marketeers of substances must check if their substances are listed in EINECS. If they are not, then they must be notified to the national competent authority in the Member State where marketing first takes place. The same rule applies to non-EU companies and importers.

After some 10 years of operation of the sixth amendment, the European Commission and Member States concluded that the system could be improved and that certain anomalies needed correction and omissions made good. Following discussions between the European Commission, Member States, and the chemical industry, the 1967 Directive was amended for the seventh time, Directive 92/32/EEC.[3] This amendment introduced amongst other things, the requirement for environmental CPL and a harmonized approach to risk assessment, the criteria *etc.* being set out in two other Directives.[4,5]

4 Existing Substances

Substances appearing on EINECS are subject now to a European Community Regulation 793/93/EEC,[6] which requires a phased collection of information on substances currently produced in quantities over 10 tonnes per annum. This is over a period of 4 years, which began in June 1994, starting with the highest tonnages first. Priority lists are made and those substances appearing on them will be evaluated for their risk to human health and the environment. Where the necessary data are not available without good reason, they will have to be generated using the same test procedures as those specified for new substances, Annex I to Directive 92/69/EEC and/or Directive 87/302/EEC.[7,8] In addition to environmental CPL, priority substances will have their potential risks assessed according to the principles of the associated EU Regulation 1488/94.[9]

The interrelationship between the various Directives/Regulations is shown diagrammatically in Figure 1.

5 Tests Required to Assess Environmental Risk

Ideally, the range of tests needed should enable effects and fate to be determined in the three environmental compartments, air, soil, and water, *e.g.*

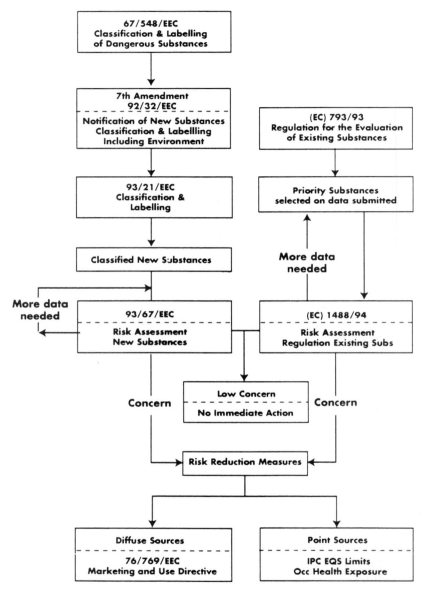

Figure 1

Air: photodegradability and transformation potential effects on the ozone layer.

Soil: effects on soil-dwelling organisms, *e.g.* earthworms, soil bacteria, higher plants, and larger animals, *e.g.* birds: fate data on degradability, adsorption/desorption for different soil types and sludge.

Water: effects on fish, crustacea, algae, higher plants; fate data on biodegradability, photodecomposition, hydrolytic stability, and effects in sediments.

The range of agreed and adopted tests is currently limited as indicated below:

Air: some methodologies exist but as yet there are no internationally agreed protocols other than procedures for estimating the potential to deplete the ozone layer and for global warming.

Soil: test procedures are available for adsorption/desorption on soils and effects on earthworms and birds. The other areas of interest are being worked on internationally, particularly the effects of chemicals on the anaerobic degradation process.

Water: short- and long-term tests are available for fish and crustacea, *e.g. Daphnia*, algae; biodegradability and stability in water (hydrolysis). Other tests are under development or the need for new ones is being considered.

The testing of 'new substances' widens in scope as the marketed quantities increase. For 'existing substances' further testing to fill in any missing data will be drawn from the tests described below, as appropriate, for risk assessment purposes. The currently available tests to be used follow closely those listed in the OECD Test Guidelines,[10] and must be carried out to the standards of Good Laboratory Practice.[11]

Base Set

For all planned marketing from 1 tonne per annum and above, the notifier must submit the following test results using the agreed methods:

Effects

- 96 h short-term fish toxicity test; the LC_{50} (dose that is lethal to 50% of test subjects) at 24, 48, 72, and 96 h and the No Observable Effects Concentration (NOEC) at 96 h.
- 48 h short-time *Daphnia* toxicity test; the EC_{50} (concentration that is effective in 50% of test subjects) at 24 and 48 h and the NOEC at 48 h.
- 72 h algal growth inhibition test.

These tests can be regarded as tests on the species in a simple aquatic food chain.

Fate

- Ready biodegradability in 28 days as a percentage of Dissolved Organic Carbon removed, min. 70%, or percentage of Theoretical Oxygen Demand, or Theoretical Carbon Dioxide, in both cases min. 60%. In each case the minimum value of the biodegradation must be achieved within 10 days of the time when the level of degradation has reached 10%.

- Bioaccumulation by determination of the value of the octanol/water partition coefficient to provide an indication of the potential to bioaccumulate.
- Adsorption/desorption screening test useful as an indicator of the potential to transfer to and from the aqueous phase.

The water solubility and the vapour pressure are also used, in conjunction with the molecular weight, as indicators of the possible distribution of the substance between water, air, and soil/sediment.

Level I. At 10 tonnes per annum or 50 tonnes in total, further testing may be required at the request of the competent authority. The choice of additional testing will reflect the outcome of any risk assessment of the substance made at the Base Set and the results of the tests obtained in the Base Set. For example, the additional tests may include those for earthworm toxicity and prolonged toxicity tests on fish and *Daphnia*. Additional biodegradation studies may be requested, especially if the initial study carried out showed a lack of ready biodegradability. A study of the hydrolysis in water as a function of pH can be asked for in this case. A test for bioaccumulation may be required if the substance has a potential to do so as evidenced by a value of the octanol/water partition coefficient exceeding 1000 (log $P \geqslant 3$).

At 100 tonnes per annum or 500 tonnes total, the competent authority is obliged to ask for the full test package. It is the responsibility of the notifier to justify why any of these additional tests need not be carried out. Any earlier assessment of risk does not significantly affect the additional testing required at this level.

Level II. At 1000 tonnes per annum or 5000 tonnes total, a programme of additional testing is required of the notifier. The tests will include studies on accumulation, degradation, and mobility, and prolonged fish toxicity testing including effects on reproduction, toxicity tests (acute and sub-acute) on birds if the bioaccumulation factor exceeds 100. Also required is an adsorption/desorption study where the substance is not particularly degradable.

Low Tonnage Chemicals. Where the planned marketing of a new substance is between 10 and 1000 kg per annum the ecotoxicological testing is very limited. None is required between 10 and 100 kg per annum, and between 100 and 1000 kg per annum only a test of degradability, although if the authorities deem it necessary, an acute toxicity test with *Daphnia* can be asked for. When the cumulative tonnage reaches 5 tonnes, the substance has to be fully tested as described above for the Base Set.

Interpreting the Data

A detailed description of the scientific interpretation of the test data is outside the scope of this chapter. Assuming that the data have been properly generated according to the principles of Good Laboratory Practice using the agreed test procedures, they are used for the purposes of classification and labelling followed by risk assessment. Currently the test data received under the notification scheme in particular are mainly those relating to the aquatic compartment.

6 Environmental Classification and Labelling

The output of the tests described in the preceding paragraphs are used to classify and label substances for their environmental effects. EU Directive 93/21/EEC sets out the principles for making these. Risk and safety phrases are used to give warnings of the environmental hazard and advice on the safe disposal of the substance. The risk phrases (R) are assigned on the basis of the lowest of the acute toxicity data determined for fish, *Daphnia*, and algae, taking into account the potential to readily biodegrade and bioaccumulation.

So far they have only been fully developed for the aquatic environment. Substances are classified as 'Dangerous for the Environment' if they have the following properties:

- R50 Very Toxic, Lowest $L(E)C_{50} \leqslant 1.0$ mg l^{-1} for fish or *Daphnia* or algae.
- R51 Toxic, as above but > 1 $L(E)C_{50} \leqslant 10$ mg l^{-1}.
- R52 Harmful, as above but > 10 $L(E)C_{50} \leqslant 100$ mg l^{-1}.
- R53 May cause long-term adverse effects in the aquatic environment. Not readily biodegradable and or log $P_{O/W}$ (octanol/water partition coefficient) $\geqslant 3.0$

The basis for the assignment of R phrases is:
R50 on toxicity alone:
R50 + R53 on toxicity with not readily biodegradable and/or log $P \geqslant 3$
R51 + R53 on toxicity with not readily biodegradable and/or log $P \geqslant 3$. At present R51 cannot be used alone for substances which are readily biodegradable and have a value of log $P \leqslant 3$.
R52 + R53 on toxicity and not readily biodegradable

The first two categories carry a symbol as well as the R phrase.

It is possible that substances assigned R52 + R53 need not be classified if there is other evidence such as a proven ability to degrade rapidly in the aquatic environment and/or an absence of *chronic* toxicity effects at a concentration of 1.0 mg l^{-1}, *e.g.* an NOEC of greater than 1.0 mg l^{-1} determined in a prolonged toxicity study with fish or *Daphnia*.

For certain substances not fitting easily into the above criteria, R52 and/or R53 alone could be applied for those where there is an indication of toxicity, persistence, potential to accumulate or the observed or predicted environmental fate may pose dangers. An example would be a poorly water soluble substance with no measured toxicity at the solubility limit but which has log $P_{O/W} \geqslant 3.0$ *and* is not readily biodegradable: this would be assigned R53 only. A measured toxicity, however, would require R50, 51, or 52 as appropriate. Again additional evidence on degradability and bioaccumulation potential may be such that the R53 classification may or may not be required. To avoid the toxicity R phrases there must be no evidence of this at the solubility limit, *e.g.* an NOEC at this limit determined in a prolonged toxicity study with fish or *Daphnia*.

The important point about the classification scheme for environmental hazard is that it is driven by toxicity not solubility.

7 Environmental Risk Characterization

New Substances

It is a requirement of EU Directive 92/32/EEC that the data submitted to the national competent authorities are such that assessment can be made of the notified substance(s) of the potential hazards to the environment and man. The data are used for making an environmental classification. A substance classified 'Dangerous for the Environment' is immediately a candidate for risk characterization. A substance not so classified may however be a candidate for assessment for environmental hazards on the basis of certain other properties, *e.g.* potential to bioaccumulate, indications of other adverse effects on the basis of toxicity testing, *etc.* The following description is concerned only with the aquatic compartment.

The EU Risk Assessment of New Substances Directive 93/67/EEC lays down the principles for making assessments. It is accompanied by technical guidance on such issues as what further testing may be indicated on the basis of the results available and when to test. Figure 2 summarizes the principles of the decision-making for the aquatic compartment.

Hazard assessment is the outcome of the iterative comparison of the predicted environmental concentration (PEC) of the substance (the 'Exposure') with a Predicted No Effect Concentration (PNEC). The latter is derived from the 'Effects' data determined in the laboratory on aquatic species by applying a safety assessment factor.

The first comparison will frequently result in a ratio PEC:PNEC > 1. The first PNEC is derived from the acute toxicity data to fish, *Daphnia*, or algae; the lowest of the three values of the $L(E)C_{50}$ is used. Before asking for further tests, refinements to the exposure, PEC, should always be made first since it is in this area that there is the highest uncertainty at this stage. To refine the estimate of PEC involves more precise knowledge of the releases during manufacture and use and how the concentration might be reduced during the progress of the substance to the receiving aquatic environment, *e.g.* a river.

Concentration reductions may arise from pre-treatment by a factory effluent treatment plant on-site: this can include dilution by other waste water arisings on-site. The discharge may be combined with the flow of municipal sewage and

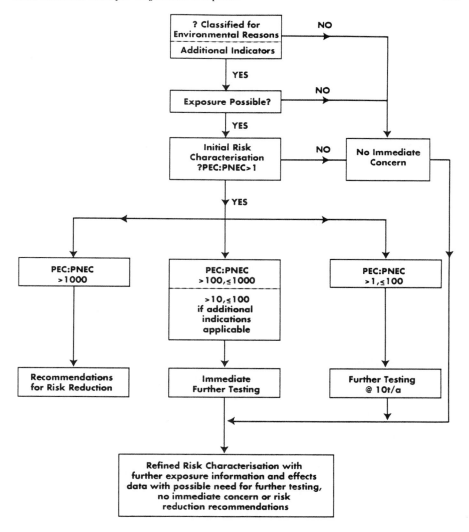

Figure 2

other waste waters and then treated in the municipal treatment works. This may well reduce the concentration of the substance still further. Finally the municipal treatment works effluent may be discharged to a river, estuary or even to coastal waters, all of which will afford dilution to a greater or lesser extent depending on, for example, flow rate relative to that of the discharge. Much of the information required will not be available in the notification dossier. This will have to be augmented by discussion between the notifier and the competent authority thus leading to at least two exposure assessments before a satisfactory estimate can be made.

Once the exposure has been agreed, the 'refined' PEC is compared with the

PNEC. If the PEC is still > PNEC, there are four alternative courses of action depending on the value of the ratio:

- PEC:PNEC > 1000; Recommendations for Risk Reduction
- PEC:PNEC < 1000, > 100 and PEC:PNEC < 100, > 10 in special cases; carry out further tests (to refine the PNEC) immediately
- PEC:PNEC < 100, > 10 and PEC:PNEC < 10, > 1 in special cases; carry out further tests (to refine the PNEC) at 10 tonnes per annum
- PEC:PNEC < 1; no immediate concern. Test at 100 tonnes per annum

The choice of further tests, essentially to refine the PNEC, will take into account what was the most sensitive of the three species tested and potential for bioaccumulation. It may be necessary at this stage to see if the PEC can or needs to be refined by carrying out further tests of degradability and/or adsorption/desorption. After this stage the refined PEC and PNEC are again compared and further testing to refine either or both may be needed. The process may be repeated linked to decisions on even more testing or risk reduction.

Assessment Factors

Mention has been made above of the need for assessment factors to estimate the PNEC. At present the factors most commonly used are somewhat arbitrary. The authorities and others characterizing risks have to try to protect whole ecosystems on the basis of the toxicity of substances determined on a few species in laboratory studies. Thus the factors tend to be very conservative to try to reduce uncertainties to a minimum. Generally the size of the factor can be decreased if longer term toxicity data are available. For instance chronic toxicity of the most sensitive species can be an order of magnitude lower than if the data were not obtained on the most sensitive species. The factor could be varied depending on whether or not the substance is discharged continuously or intermittently: the factor in the latter case might be lower than for the former simply because the receiving water has more chance to effectively reduce concentration by dilution.

The current factors used are applied to the toxicity data as follows:

- 1000 to the lowest $L(E)C_{50}$ values for fish, *Daphnia*, or algae
- 50 to long-term toxicity data from two species in two taxonomic groups: applied to the lower NOEC
- 10 to long-term toxicity data from fish, *Daphnia*, or algae: applied to the lowest NOEC

If field data exist, the factors are reviewed on a case-by-case basis.

Existing Substances

The environmental risk assessment of existing substances is made on those on a priority list. The outcomes of the EU Risk Assessment Regulation for these substances are as follows:

- on the available information the substance is not considered to require further testing and is of no immediate concern.
- there is a need for further information to characterize the risk.
- the substance is of concern and should be subject to risk reduction measures.

These conclusions are not dissimilar to those for new substances, which enter the assessment process depending on whether or not they are environmentally classified. The actual process of assessment for both is iterative.

Technical guidance documents for use in making risk assessments of new and existing substances according to the principles laid down in their respective risk assessment directive and regulation are available.

The Use of Risk Assessment for Control Measures

The acquisition of data on the effects and fate of chemicals together with information on the estimated or measured environmental concentration to characterize potential risk is now a prerequisite for proposals for control measures. Within the EU legislative framework there are facilities for a Member State to propose control measures for a substance that it considers harmful. To do this the Commission has to be notified of the draft proposal according to the requirements of EU Directives 83/189/EEC and 94/10/EEC.[12,13] The basis for the proposal must be clearly set out and a risk assessment included, if appropriate (EU Directive 94/10/EEC). Unless there are compelling and urgent reasons for immediate control measures to be put in place, the proposer cannot implement his controls for 3 months. During this period other Member States are consulted and if there is no objection from them and/or the Commission the measures can be put in place.

If a detailed opinion is given by one or more Member State and/or the European Commission within the 3 month standstill period, to the effect that the proposal should be amended to eliminate or reduce any barrier to trade, the proposer has to report to the Commission what action it is intended to take on the detailed opinion. The introduction of the proposed measure may be made 6 months after the proposal was notified. If the European Commission within the 3 month standstill period gives notice of its intention to adopt a directive on the subject, the originator of the proposal may not implement it for 12 months from its notification. If the EU Council of Ministers is already considering a directive on the same subject, then the proposed regulation cannot be adopted until 12 months following the date of submission of the proposal.

There is another EU Directive, 76/769/EEC, the 'Marketing and Use' Directive,[14] which is used for European Community-wide control measures. This is a framework directive which is amended every time one or more substances are to be subjected to EU-wide control measures. Frequently substances notified under the EU Technical Standards Directives, 83/189/EEC *etc.*, find their way onto the list of controlled substances. The measures may range, for example, from concentration limits in preparations and discharges and changes in classification and labelling, to an outright ban on particular uses of a substance.

References

1 67/548/EEC, 'Parent directive harmonising the requirements for the Classification, Packaging and Labelling of Dangerous Substances', *Off. J.*, 1967, **10**, No. 196, 267 *et seq.*
2 79/831/EEC, '6th Amendment of 67/548/EEC, introducing a notification scheme for new substances', *Off. J.*, 1979, **22**, No. L259, 10 *et seq.*
3 92/32/EEC, '7th Amendment of 67/548/EEC. Notification of new substances together with environmental classification and labelling and risk assessment requirements', *Off. J.*, 1992, **35**, No. L154, 1 *et seq.*
4 93/21/EEC, 'Labelling guide: gives criteria for selecting risk and safety phrases', *Off. J.*, 1993, **36**, No. L110 A, 1 *et seq.*
5 93/67/EEC, 'Principles of risk assessment for new chemicals', *Off. J.*, 1993, **36**, No. L227, 9 *et seq.*
6 793/93/EEC, 'Regulation requiring the evaluation and control of the risks of existing substances', *Off. J.*, 1993, **36**, No. L84, 1 *et seq.*
7 87/302/EE, 'Compendium of test methods – Level I/II', *Off. J.*, 1988, **31**, No. L133, 1 *et seq.*
8 92/69/EEC, 'Compendium of test methods – Base Set', *Off. J.*, 1992, **35**, No. L383A, 1 *et seq.* (This is the revision of the earlier Directive 84/449/EEC.)
9 1488/94, 'Regulation laying down principles for the assessment of risks for existing substances', *Off. J.*, 1994, **37**, No. L161, 3 *et seq.*
10 OECD Guidelines for the Testing of Chemicals, vols. I and II, OECD, Paris, 1993.
11 87/18/EEC, 'Directive requiring the application of the principles of Good Laboratory Practice to the testing of chemicals', *Off. J.*, 1987, **30**, No. L15, 29 *et seq.*
12 83/189/EEC, 'Technical Standards Directive laying down rules for the notification of proposals by Member States for control measures', *Off. J.*, 1983, **26**, No. L109, 8 *et seq.*
13 94/10/EEC, 'Amendment of 83/189/EEC requiring risk assessment in support of control measures', *Off. J.*, 1994, **37**, No. L109, 37 *et seq.*
14 76/769/EEC, 'Framework directive for restrictions on the Marketing and Use of Dangerous Substances and Preparations. Amended as required as new, to this Directive, substances are to be subject to control measures across the European Union', *Off. J.*, 1989, **32**, No. L398, 19–23.

Chapter 14

Environmental Classification and Risk Assessment

By Derek Brown

CONSULTANT, 26 BRUNEL ROAD, PAIGNTON TQ4 6HN, UK

1 Introduction

The chemical industry and legislators, concerned with the sale of chemicals in respect of their possible environmental impact, have three common objectives. The first objective is to ensure that substances whose intrinsic properties may be hazardous to the environment are identified. The second objective is to ensure that the risks posed by the use of such substances are evaluated. The third objective is to ensure that substances with hazardous properties are used in a way that does not pose unreasonable risks to the environment. This chapter considers the legislation in the European Union (EU), which has been framed to meet these objectives.

2 Identification of Hazardous Properties (Classification)

Classification Packaging and Labelling of Dangerous Substances Directive

The identification of substances with intrinsically hazardous properties, whether by reason of physical properties such as flammability or explosivity, potential adverse human health effects (toxicity, carcinogenicity, sensitization, *etc.*), or adverse effects on the environment, falls under the classification part of the Classification, Packaging, and Labelling of Dangerous Substances Directive (the CPL Directive).

The CPL Directive dates from 1967 (Council Directive 67/548/EEC) but has been subject to a number of major updates and extensions either via Council Amendments (most recently the 7th Amendment: Council Directive 92/32/EEC) or by Commission Directives that 'Adapt to Technical Progress' a whole set of Annexes to the Directive. This Directive and all its amendments and adaptations are enacted into the laws of the EU Member States via national legislation which in the UK is primarily contained within the CHIP (Chemicals Hazard Information and Packaging) and NONS (Notification of New Substances) Regulations.

6th Amendment of CPL Directive

The 6th Amendment of the CPL Directive (Directive 79/831/EEC) introduced three important concepts from the viewpoint of the environment and the classification and risk assessment process.

The first, for which this amendment is probably most well known, is the requirement that 'new' substances placed on the EU market are subject to a formal notification scheme. This notification includes within its provisions the requirement that a series of prescribed studies (the 'base set'), aimed at identifying physical, human health, or environmental hazards, must be carried out and reported before the substance can be sold. 'New' in this context means not on the European Inventory of Existing Commercial Chemical Substances (EINICS). EINICS was open for the inclusion of those substances that were marketed for commercial purposes between 1st January 1971 and 18th September 1981 and is now closed. Under both the 6th and 7th Amendments to the CPL Directive there are provisions for further testing of new substances at higher tonnages (levels 1 and 2).

The second concept in the 6th Amendment of the Directive was the introduction of the 'Dangerous for the Environment' hazard category in Article 2.2(k). However, at the time of the 6th Amendment no criteria to use this classification were laid down.

The third concept, which at the time received little attention and no enabling action, was the requirement under Article 3.2 for environmental hazard assessment.

Development of Classification Criteria under the '6th Amendment' Directive

Article 2.2 (k) of the 6th Amendment Directive defines substances and preparations as 'Dangerous for the Environment' when their 'use presents or may present immediate or delayed risks for the environment'.

In the discussions on the criteria for the dangerous for the environment classification, the Commission made clear that the criteria should, at present, be confined to substances and in principle use 'base set' data. Furthermore, the use of the substance should not be considered, as this is part of the risk assessment process. In respect of the latter point, the corresponding article, 2.2(o), in the 7th Amendment omits the words "the use of which". This article makes other modifications to the 6th Amendment text and reads as follows: "Substances and preparations which, were they to enter the environment, would present or may present an immediate or delayed danger for one or more components of the environment".

The proposed use of 'base set' data posed the problem as to how to reconcile limited data applicable primarily to the aquatic environment with the complexity of the environment and its ecosystems. The criteria also needed to encompass the concept of 'immediate or delayed danger' and to achieve a proper balance of sufficient classification without over-classification. A further problem was that of poorly water-soluble substances, some of which may have long-term effects not manifest in the short-term acute studies of the 'base set'.

The 18th Adaptation to Technical Progress (ATP): Classification as 'Dangerous for the Aquatic Environment'

The results of the discussions on the environmental classification criteria for substances outlined above are to be found as part of the 18th ATP published as annexes to the Commission Directive 93/21/EEC in the *Official Journal* (L110A, 4 May 1993). The 18th ATP also includes the rules for the labelling of classified substances. The earlier 12th ATP also included environmental classification and labelling but in this respect has been entirely superseded by the 18th ATP.

The 18th ATP defines a number of 'R' phrases (so-called 'risk' phrases) predominantly relevant to aquatic organisms and some 'safety net' 'R' phrases to enable classification of other substances to be made on a case-by-case basis.

The toxicity criteria that drive the 'R' phrases are based on the 96 h LC_{50} (concentration that is lethal in 50% of test subjects) to fish, the 48 h EC_{50} (concentration that has an effect in 50% of test subjects) to *Daphnia*, and the 72 h IC_{50} (concentration giving 50% inhibition) for algae, and the lowest of these three values (or the lowest of the available data) is taken for classification purposes. Where there are conflicting data for a given end-point, say fish LC_{50}, scientific judgement may be used to select the LC_{50} value to be used for classification.

The potential for 'delayed danger' is defined in terms of degradability and bioaccumulation potential. 'Readily degradable' is defined primarily in terms of the results of a ready biodegradability study and is looking for ultimate degradation as the end-point.

Bioaccumulation is defined in terms of the log P_{ow} (octanol/water partition coefficient). It is recognized that log P_{ow} is a surrogate for actual bioaccumulation and may be overridden by an experimentally determined BCF (Bioconcentration Factor).

There are six aquatic environment 'R' phrases in the 18th ATP as follows:

- R50:   Very toxic to aquatic organisms
- R50/53: Very toxic to aquatic organisms, may cause long-term adverse effects in the aquatic environment
- R51/53: Toxic to aquatic organisms, may cause long-term adverse effects in the aquatic environment
- R52/53: Harmful to aquatic organisms, may cause long-term adverse effects in the aquatic environment
- R52:   Harmful to aquatic organisms
- R53:   May cause long-term adverse effects in the aquatic environment

The various criteria which drive the risk phrases may be summarized in a simplified form as follows:

- R50:   Toxicity at or below $1 \, \text{mg} \, l^{-1}$.
- R50/53: Toxicity at or below $1 \, \text{mg} \, l^{-1}$ and in addition the substance must be either not readily degradable and/or bioaccumulative.
- R51/53: Toxicity above $1 \, \text{mg} \, l^{-1}$ and at or below $10 \, \text{mg} \, l^{-1}$ and in addition the substance must be either not readily degradable and/or bioaccumulative. (Note that R51 cannot be used by itself!)
- R52/53: Toxicity above $10 \, \text{mg} \, l^{-1}$ and at or below $100 \, \text{mg} \, l^{-1}$, and in addition the substance must be not readily degradable.
- R52:   'Safety net' clause with no specific criteria. The guidance given by the 18th ATP is that it should be used for substances not otherwise classified for the environment, "but which on the basis of the available evidence concerning their toxicity may nonetheless present a danger to the structure and/or functioning of aquatic ecosystems". This phrase in practice has found little or no application.
- R53:   'Safety net' clause. 18th ATP guidance specifies its use for substances "which on the basis of the available evidence concerning their persistence, potential to accumulate, and predicted or observed environmental fate and behaviour may nevertheless present a long-term and/or delayed danger to the structure and/or functioning of aquatic ecosystems". This R53 phrase unlike the R52 phrase is exemplified in the 18th ATP for sparingly soluble substances (solubility $\leq 1 \, \text{mg} \, l^{-1}$) and has found widespread utility for such substances.

It should be noted that the R52/53 and the R53 'R' phrase criteria have 'escape clauses'. These allow otherwise classified substances to be exempt if they can be shown by additional degradation or toxicity studies not to present a potential long-term danger to the aquatic environment.

18th ATP: Classification for the Non-aquatic Environment

The 18th ATP has a further six 'R' phrases (R54–R59) that are intended for use opposite the non-aquatic compartments of the environment. Apart from R59, there are no criteria as yet defined for them and, until such time as criteria may be developed, they should not be used:

- R54: Toxic to flora
- R55: Toxic to fauna
- R56: Toxic to soil organisms
- R57: Toxic to bees
- R58: May cause long-term effects in the environment; 'safety net' phrase
- R59: Dangerous for the ozone layer

The R59 classification is made primarily on the basis of substances listed in Annex I of the Council Regulation EEC No. 549/91 on substances that deplete the ozone layer. Substances in Groups I–V of this Annex are classified as R59 with the environmental symbol N (see below) and those in Annex VI as R59 alone.

18th ATP Labelling

The 18th ATP specifies both the criteria for classification and the resulting label that classified substances should carry.

For substances classified R50, R50/53, R51/53, or R59 (plus 'N') the environmental symbol 'N' (dead fish and tree) together with the warning words 'Dangerous for the environment' is required as part of the label. For all classified substances the appropriate 'R' phrase is required together with an 'S' (Safety Advice) phrase.

The 18th ATP has a choice of possible environmental 'S' phrases as follows:

- S35: This material and its container must be disposed of in a safe way
- S56: Dispose of this material and its container to hazardous or special waste collection point
- S57: Use appropriate containment to avoid environmental contamination
- S59: Refer to manufacture/supplier for information on recovery/recycling; used for substances classified opposite the ozone layer
- S60: This material and its container must be disposed of as hazardous waste
- S61: Avoid release to the environment. Refer to special instructions/safety data sheet.

The ATP working party (see below), responsible for proposing the environmental classification of Annex I substances, has adopted a normal working practice of giving all substances classified in respect of the aquatic environment the S61 phrase and in addition for the R50/53 substances the S60 phrase. R59 substances will receive the S59 phrase.

The Classification Process in the EU

The classification procedure may be undertaken in one of three ways.

Annex I Substances. If a substance is already on Annex I of the CPL Directive by reason of another dangerous property, the environmental classification process is carried out by the ATP process. This is currently underway and involves a working party of the Commission and Member States, with industry observers. The working party is reviewing all of Annex I (about 1800 substances) and preparing appropriate classification proposals which, if adopted by the ATP Technical Progress Committee (TPC), will be incorporated in Annex I and published as an ATP in the Official Journal. A similar process is undertaken for existing substances which are newly proposed for inclusion into Annex I by reason of one or more dangerous properties.

Newly Notified Substances. Where a substance is newly notified, with a full notification as required under the 6th and 7th Amendments of the CPL Directive, its total classification, including the environmental classification, is automatically considered under a similar ATP process to that described for Annex I substances.

The only difference is that, for reasons of confidentiality, there are no industry observers in the working party. Where newly notified substances are classified they are automatically included in Annex I.

Substances Not on Annex I.   For all substances that are on EINECS but are not in Annex I, article 6 of the 7th Amendment places an obligation on manufacturers, distributors, and importers to "make themselves aware of relevant and accessible data" and "on the basis of this information, package and provisionally label the substance according to the rules laid down...". This provisional labelling is often termed 'self-classification' and, as environmental classification is a complicated area, CEFIC have attempted to help those responsible by issuing a guidance document.

Significance of Environmental Classification

As with all such classifications in the EU, environmental classification is a warning of intrinsic hazard and is not based on a risk assessment. The classification and associated 'R' and 'S' phrases relate to the substance as supplied and "not necessarily in any different form in which they may finally be used, *e.g.* diluted" (quote from the general introduction to labelling in the 18th ATP).

The environmental classification scheme is only now being introduced into the national legislation of the EU Member States. The requirement to classify and label for environmental effects applies only to substances and, as yet, not to preparations containing those substances. In consequence there are currently very few containers carrying the environmental symbol and 'R' and 'S' phrases. Estimates of the number of substances likely to be classified as 'Dangerous for the Environment' are approximately 50% of all substances, of which about half are likely to require the environmental symbol and warning.

Environmental Classification and Labelling of Preparations

The classification, packaging, and labelling of preparations is covered by the Classification, Packaging and Labelling of Dangerous Preparations Directive 88/379/EEC. Proposals to amend this Directive are currently being prepared by the Commission and should be submitted to Council and Parliament during 1996. The proposed amendments include the 'Dangerous for the Environment' classification for preparations and extension of the scope of the Directive to include pesticide preparations. The probable time frame for the adoption of the amended Directive is by 1998/1999, with incorporation into national legislation by 1999/2000.

The classification process within the Preparation Directive predominantly relies on standard 'concentration limits'. However, the proposals provide for individual concentration limits to be adjusted down if a substance is particularly toxic, and this is likely to operate for the 'Dangerous for the Environment' classification if the aquatic toxicity of a substance is $<0.1 \, \text{mg} \, \text{l}^{-1}$. Similarly, there are proposed provisions for the aquatic toxicity of the preparation to be tested if the concentration limit approach appears to be giving an incorrect classification.

The likely Commission proposals for the standard concentration limits for the 'Dangerous for the Environment' (aquatic environment) classification of preparations are 25%, 2.5%, and 0.25%. The application of these concentration limits is complicated but may be summarized in a simplified form as follows:

- Preparations containing >25% of R50 substances will be classified R50: if <25% will not be classified
- Preparations containing >25% R50/53 substances will be classified R50/53: if 2.5–25%: R51/53; if 0.25–2.5%: R52/53
- Preparations containing >25% R51/53 substances will be classified R51/53: if 2.5–25%: R52/53
- Preparations containing >25% R52/53 substances will be classified R52/53

For preparations containing substances that are dangerous for the ozone layer the concentration limit for NR59 (classified with the symbol) or R59 is 0.5%.

The Commission proposals for the amendment to the Preparations Directive are also likely to include some modifications to the 'R' and 'S' phrases for substances as follows:

- R50/53: Contains substances very dangerous for the aquatic environment
- R50 or R51/53: Contains substances dangerous for the aquatic environment
- R52/53, R52 or R53: Contains substances harmful to the aquatic environment
- NR59 or R59: Contains substances depleting the ozone layer
- Additional 'S' phrase: "For use and disposal follow manufacturer's instructions"

3 Environmental Risk Assessment

The second common objective identified in the introductory paragraph to this chapter was to ensure that the risks posed by substances with hazardous properties are evaluated. Two legislative instruments have been enacted for this purpose within the EU depending on whether the substance is 'new' (not in EINECS) or 'existing' (in EINECS).

7th Amendment of CPL Directive

Under the 7th Amendment to the CPL Directive (92/32/EEC) the requirement for newly notified substances to be subject to a formal risk assessment process was introduced. This risk assessment covers human health, physical hazards, and risk to the environment posed by the substance in both its manufacture and subsequent use. The risk assessment is formally prepared by the competent authority to whom the notification is made on behalf of the whole EU. The Commission and other member states have the opportunity to comment on the risk assessment. The notifier, when making the notification, may submit his own risk assessment. Within the UK a positive encouragement to do this is given by the competent authority which charges a fee if the notifier does not submit an acceptable risk assessment.

Directive on Risk Assessment of Notified Substances

The general principles for assessing the risks of notified substances are given in Directive 93/67/EEC. This is supplemented by a rather large Technical Guidance Document which "represents a consensus between Member States and DG XI", but "is not legally binding and should be reviewed in the light of the assessor's professional judgement".

Environmental risk assessment is an exceedingly complex topic in that in principle all environmental compartments, all species, and all life stages need to be considered. In practice, most experimental data on effects relate to the aquatic environment. There may be some data relevant to soil, but with the exception of some knowledge of effects on atmospheric chemistry, almost none relevant to air. This is reflected in the technical guidance where much of the guidance is devoted to the aquatic environment.

The basis of environmental risk assessment is the PEC/PNEC ratio. PEC is the Predicted Environmental Concentration and PNEC is the Predicted No Effect Concentration. Where the PEC/PNEC is less than 1, then the competent authority will conclude that there is no immediate concern and no action is necessary until the next tonnage level trigger is reached. At this time, the risk assessment will be reviewed in the light of the new tonnage and any further information.

Where PEC/PNEC exceeds 1, three courses of action are open to the competent authority. The first is to postpone a decision until the next tonnage trigger. The second is to request immediate further testing to refine the PEC and/or PNEC estimates. The third is to propose some form of risk reduction that will reduce the exposure to the substance. Guidance as to which option should be selected is given in the Risk Assessment Directive in terms of the PEC/PNEC ratio, but some discretion is given to the competent authority in coming to a decision.

When the PEC/PNEC ratio is found by the competent authority to exceed 1, the notifier may be informed and given the opportunity to comment. This is the normal practice within the UK.

Environmental Risk Assessment in the Aquatic Compartment

For a newly notified substance the base set of data is used to assess PEC with the main inputs being tonnage, use, partitioning behaviour, and degradability. Certain generic discharge and sewage treatment efficiency scenarios are laid out in the technical guidance and various models are employed in the calculation.

If a notifier believes that an alternative scenario is more appropriate for the substance being notified this case can be made as part of the notifier's risk assessment and/or response to the competent authority's risk assessment. Such an exposure scenario might well include a consideration of the time and distance variation in the exposure arising from, for example, an intermittent discharge of the substance into a stream or river. Thus available information on patterns of manufacture and likely use scenarios will be extremely useful in estimating realistic PEC values. Measured concentrations of the substance or a close analogue in realistic manufacturing and/or use situations will be of even greater value.

The base set of acute aquatic data is used to assess the PNEC. This is done by taking the lowest acute toxicity figure (fish 96 h LC_{50}, *Daphnia* 48 h EC_{50}, algae 72 h IC_{50}) and applying an assessment factor to give the PNEC. As indicated above the risk assessment needs to consider all species and all stages in the life cycle, and in consequence a relatively conservative factor of 1000 is normally used. Where the exposure (PEC) scenario is indicating significant time/distance variation, there may be an argument for considering a lowering of this assessment factor.

The PEC/PNEC figure derived from the above is regarded as a crude initial PEC/PNEC and 'risk reduction' will rarely, if ever, be proposed on this basis. However, where the initial PEC/PNEC ratio gives cause for concern, the second stage of the risk assessment will involve a refinement of the PEC and/or the PNEC.

PEC refinement could involve further laboratory studies on degradability and/or partitioning behaviour. It could also involve analytical monitoring of actual discharge and/or environmental concentrations.

PNEC refinement involves further aquatic toxicity testing using longer term chronic studies. If results from chronic studies on two taxonomic groups (usually two of fish, *Daphnia* or algae) are available then an application factor of 50 is normally applied to the lower NOEC (No Observed Effect Concentration). If three chronic studies are available then a factor of 10 is normally applied to the lowest NOEC.

The PEC/PNEC refinement may in principle go through a number of cycles and may also include more sophisticated mesocosm toxicity studies and/or ecological observations in the receiving environment (more relevant to high tonnage existing chemicals). Only if this process is still indicating a ratio greater than unity may risk reduction measures be proposed.

Existing Chemicals

Somewhat in parallel with the provisions of the CPL Directive on the notification and risk assessment of new substances is the Council Regulation No. 793/93 on the Evaluation and Control of Existing Chemicals (the Existing Chemicals Regulation). This sets out a framework for the reporting of known data and classification proposals for existing substances, a procedure for deciding which substances should be given priority for risk assessment, and a risk assessment procedure. The regulation, unlike a directive, has immediate effect in all member states. It is seen as complementary to the long-standing OECD HPV (High Production Volume) Chemicals programme.

The reporting requirements under the existing chemicals regulation apply to substances listed on EINECS and are phased according to tonnage/manufacturer or importer:

- Tonnage > 1000 tonnes per annum per company and on Annex I of Regulation by 4th June 1994
- Tonnage > 1000 tonnes per annum per company and not on Annex I by 4th June 1995
- Tonnage 10–1000 tonnes per annum per company between 4th June 1996 and 4th June 1998

The reporting is done via an electronic format known as HEDSET. The first part, giving tonnages and uses, must be filled in by all manufacturers and importers who make/import the substance in question above the tonnage threshold. The second part, giving the required known physical, toxicological, and environmental data plus classification proposals, may be filled in by an individual manufacturer/importer. Where a substance is manufactured/imported by a number of different companies there is encouragement for one company to fill in this second part on behalf of all the submitting companies. There is a much reduced data requirement for the lower tonnage materials.

Priority Setting and Risk Assessment

Following the submission of the HEDSET data, a priority list of substances for evaluation is drawn up by inputting the data into the IPS (Initial Priority Setting Scheme) and by consulting member states on their priority lists. The objective is to have a succession of priority lists each containing about 50 substances, each new list being published 1 to 2 years after the preceding one. The priority lists are published in the Official Journal. The first priority list has been published and, at the time of writing, the second is being actively considered.

For each substance on the priority list a member state is appointed as rapporteur responsible for carrying out the risk assessment. The manufacturers/importers who have submitted HEDSET data on a priority list substance must forward all relevant information and corresponding reports within 6 months of publication to the rapporteur. If base set information is missing, this must be obtained and the test reports submitted within 12 months unless the rapporteur agrees that this is not necessary or not possible. The rapporteur may also recommend that further data are obtained and the regulation lays down a formal procedure for considering this recommendation. The rapporteur having completed the risk assessment may suggest a strategy for limiting the risks. If this includes proposals to limit marketing and use, the rapporteur must submit an analysis of advantages and drawbacks to the substance and the availability of replacement substances.

In parallel with the risk assessment of notified substances under the 7th Amendment Directive there is a Commission Regulation (1488/94) on the principles for the assessment of risks to man and the environment of existing substances, and this is backed by a Technical Guidance Document with the same status as for new substances'. Although the Technical Guidance Documents for new and existing substances are currently different, the general principles are the same, and the Commission has proposed that the two documents should be combined. It is also accepted that risk assessment practice is at a formative stage and that all concerned need to "learn by doing".

4 Risk Management

The third common objective, indicated in the Introduction, was to ensure that substances with hazardous properties are used in a way that does not pose unreasonable risks to the environment (risk management).

One of the basic principles of the CPL Directive is 'risk management'. The labelling of classified substances to warn users of the hazardous property is part of this strategy, as are the provisions for specialized packaging (*e.g.* child-proof closures) for certain classified materials. A further requirement introduced in the 7th Amendment is that a safety data sheet must be supplied to the professional users of dangerous (classified) substances and that this data sheet must contain the information "necessary for the protection of man and the environment". The risk assessment process and possible risk reduction measures, also introduced with the 7th Amendment, are also clearly part of 'risk management'.

From formal risk assessments carried out under the notification procedures for new chemicals or on priority substances under the Existing Chemicals Regulation, there may be a conclusion from the refined PEC/PNEC ratio that risk reduction is necessary.

There is little experience of the type of risk reduction measures that may be proposed in this legislative area. These could include a prescribed treatment system for facilities using the substance, limitations on quantities used in specified time periods, limitations on the mode of use of the substance, limitations on the disposal of the substance, recycling requirements, or an outright ban. These measures all affect the PEC and need to be considered by both industry and the legislator as an on-going part of the risk assessment process rather than as a final stage.

The legislative instrument for enforcing risk reduction measures is the Marketing and Use Directive (76/769/EEC). However, it is to be hoped that it will not be necessary to invoke formal legislation for most situations.

5 Conclusions

This chapter has tried to show that there should be a common agenda between those responsible for legislation and the chemical industry in the overall area of risk management, and three key activities have been identified.

Hazard identification and warning (classification and labelling) is the first element of risk management. This is a relatively simple and limited hazard identification based on intrinsic properties and is not a risk assessment.

Risk assessment is the second element and in the area of environmental risk assessment the calculation of 'safe' levels relies on largely empirical assessment factors. These need to be used in this context and note taken of the pattern of use and quantity of the substance under consideration in coming to conclusions based on these factors. Similarly exposure calculations and models are useful tools but their conclusions need to be considered alongside actual measurements of the substance or of similar substances in the environment. The overall conclusions of a theoretical environmental risk assessment should, wherever possible, be checked opposite what is known in the real environment.

Exposure reduction is the third element of risk management and is sometimes seen as the totality of risk management to be considered as a final stage. In fact exposure reduction, which is the only way to reduce risk, is an intrinsic part of the total risk management process. As such, exposure reduction needs to be factored in at all stages of the risk assessment process.

Chapter 15

Communicating the Results of a Risk Assessment: Lessons from Radioactive Waste Disposal

By Ray Kemp

ENTEC UK LTD, 17 ANGEL GATE, LONDON EC1V 2PT, UK

1 Introduction

Radioactive waste disposal has become one of the most publicly controversial areas of environmental policy implementation over recent years. As a result, risk communication in this field has reached a high level of sophistication and it can provide useful lessons in the communication of the risks from chemical pollution.

Technical assessments have been extensive, and international co-operation between the nuclear industry/developers and between various regulatory organizations has been extensive. Given the long time frames with which the radiological assessments deal, environmental and human safety issues are often addressed in terms of *risk*. 'Risk assessment' is employed here as a generic term to include a broad range of techniques including probabilistic safety assessments, probabilistic systems assessments, quantitative risk assessments, post-closure safety studies, and so on. In essence, risk assessments are employed by developers and by regulators alike, in order to take account of the large-scale uncertainties and the elements of subjective judgement that are intrinsic to such long-term forecasting problems. The concept of risk has become an essential 'tool of the trade'.

At the same time, those who undertake risk assessments in relation to radioactive waste disposal issues have become acutely aware of two complicating factors. First, that their risk assessments must at some stage be *communicable* to others, in particular to decision-takers, regarding the acceptability or otherwise of radioactive waste disposal proposals. Secondly, because of the controversial nature of such projects, the decision-making process in democratic societies invariably includes a public element where the proposals must be explained and assessments made explicit in some form of public arena.

This chapter will introduce the concept of *Paradigms of Risk*, explaining the changing concepts of risk and of risk perception. Next, there will be a discussion of the concepts of 'acceptable' or 'tolerable risk' and of how these concepts enter the decision-making arena. Thirdly, the chapter instigates a discussion of approaches to risk communication and questions of public trust. Fourthly, we raise lessons learnt from recent experience. Issues of procedure are also raised, in particular highlighting the need for accessibility to risk information and associated problems with procedure, such as formal inquiries, the roles of independent expert groups, informal procedures, and other potential alternative processes. The chapter concludes with a discussion of key principles and of practical considerations to be borne in mind by the regulator when communicating information about the risk assessment process.

2   The Paradigms of Risk

It is apparent that distinct areas of risk study have emerged over recent years. To the extent that these distinct areas of science appear to employ different concepts or explanations of the term 'risk', utilize different methods of data gathering and analysis, and hold to different epistemologies, they may be said to constitute different *paradigms* (or pre-paradigms). Thomas Kuhn defined paradigms as distinct fields and methods of research split into defined problems that attract a distinct group of adherents (see reference 1, p. 10).

No overall consensus has yet emerged between these differing risk paradigms, but each is troubled and challenged by the insights and work of the others. Concomitantly, no single paradigm provides a range of insights and tools that is sufficiently comprehensive to make the alternative approaches redundant. Rather, each approach builds upon certain strengths and challenges certain weaknesses of its competitors; they are co-dependent and interdependent.

What I shall call the *first paradigm of risk* developed from the 1960s onwards with the aim of providing more accurate predictions of the reliability and consequences of failure associated with the aircraft and chemical industries. Following the events at Three Mile Island in 1979, there was increased impetus towards the application of Probabilistic Risk Assessment within the nuclear power industry, in the search for predictive tools to assess the level of safety associated with particular industrial plant or processes. This first paradigm of risk relies upon historical data on plant performance and known event sequences, probability estimates and consequence analysis, to provide a technical assessment of the likely harm associated with a given (existing or proposed) activity. Risk within this paradigm is defined essentially as a product of probability × consequences (see Table 1).

The *second paradigm of risk* followed on from the first paradigm's concern for plant safety when questions began to be asked about what level of safety was sufficient. In 1969, Chauncey Starr began to raise the question "how safe is safe enough?". This was the beginning of the concern for developing standards of acceptable risk whereby a balance is sought to be achieved between the evaluation of the likelihood of an adverse event or accident occurring and the likely consequences, as against the appropriate levels of safety investment and regulatory control necessary to make that risk appear acceptable.

Table 1   *Four paradigms of risk*

First paradigm	Probabilistic Risk Assessment 'risk = probability × consequences'	• technical–scientific basis • empirical basis • probabilities × consequences • uncertainties
Second paradigm	Acceptable Risk 'acceptable risk = risk/benefit trade off'	• comparative risk assessment • perceived risk • expressed and revealed preferences • regulatory criteria • empirical and normative basis
Third paradigm	Cultural Context of Risk 'risk = culturally laden phenomenon'	• social–psychological and anthropological basis of concern • perceived risk • cultural and organizational typologies • empirical and normative basis
Fourth paradigm	Risk Communication 'risk = environmental hazards open to distortion through processes of communication'	• social amplification and social attenuation of risk • actual and perceived risks • empirical and normative basis

Thus this second paradigm adds a normative element to the empirical basis of the first paradigm of risk. Now the focus is on the introduction of standards regarding the level of risk that is acceptable, both to individuals and to society. The fact that a risk is accepted historically does not, of course, mean that it is acceptable.[2] For example, the empirically demonstrable level of fatalities due to air traffic incidents indicates that the risk to the individual, of death from an aircraft crash, is approximately one in a million per year. This does not mean of course that this is an acceptable level, as the Royal Society Study Group on Risk Assessment put it:

"The fact that a risk is accepted is by no means a guarantee of its acceptability. Death has to be accepted by the individual, but is not acceptable to him. Moreover, acceptance may be associated with ignorance or a simple disbelief in the existence of any significant risk".[2]

As far as the regulator is concerned, this raises the basis for the ALARA/ALARP (as low as reasonably achievable/practicable) principles, which are now employed in health and safety and environmental pollution control decisions in the UK. Following the Sizewell Inquiry into proposals to build a Pressurized Water Reactor nuclear power station at Sizewell in Suffolk, UK, there has been much discussion of the concept of 'tolerable' as opposed to 'acceptable' risk. The inquiry inspector, Sir Frank Layfield, introduced the notion that many people will 'tolerate', not accept a risk, because they may see certain benefits associated with that risk and are reassured that it is properly and openly regulated (see references 3 and 4).[3,4]

A *third paradigm of risk* may be seen to have arisen from wider social science considerations of the cultural context within which risk-related decisions are made. Studies by social psychologists of the differences between various individuals and different groups in their perceptions of risk and of risky activities were important in influencing the debate about acceptable levels of risk. The literature is now extensive. More recently, there has been a development to examine the cultural influences upon risk perceptions, taking the lead from the work of social anthropologists such as Mary Douglas and Michael Thompson. This perspective sees the distribution of risk in society as culturally laden as a consequence of decisions taken in the past, and of the form and structure of the organizations which made them. A distinct but related strand of work has come to examine the processes of social stigma and public trust. These are said to influence individual and group perceptions of risk. Cultural and organizational factors which help to create social stigma around groups and activities affected by environmental risk are analysed in terms of social cohesion and processes of communication.

A *fourth paradigm of risk* evolves from the foregoing by concentrating upon questions of risk communication. The visibility of certain processes such as radioactive waste disposal, incineration of toxic wastes, and the development of large industrial plant, for example, and the growth of public concern for their own well-being and for the health of the environment, has led to evolving work on risk communication, *i.e.* how information about risky activities is structured and transposed in the decision-making process.

There has been a growing concern with the strength of public opposition to the siting of potentially hazardous activities and this led Kasperson *et al.*,[5] to present their 'fledgling conceptual framework' on risk communication. Their social amplification of risk model sets out the main factors that influence individual and societal judgements about risk events and that bring about the divergence of perception between technical expertise and the general public of the level of risk associated with a particular activity. In other words, this paradigm seeks to explain how the first paradigm of risk becomes distorted by the second and third paradigms. Risk can then be overplayed, overestimated, overemphasized, and blown out of all proportion; *i.e.* amplified. Conversely, risks may be underestimated, underemphasized, and improperly explained to the community at large; *i.e.* attenuated.

It is quite clear that the adversarial nature of the UK planning process provides the very arena within which risk communication on a complex technical matter, such as the long-term radiological assessment of the deep disposal of radioactive waste, may be subject to competing influences upon the information to be communicated. This is the very kind of arena whereby the 'innocence' or 'guilt' of the messenger may be readily influenced by the acceptability of the message being brought. The challenge which this fourth paradigm therefore addresses is the efficacy of alternative decision-making procedures by which the public may participate in resolving risk issues that directly or indirectly affect their lives.

3 Factors Influencing the Public Acceptance of Risk

In the USA in 1978, Fischhoff *et al.*[6] set out to examine the differences between expert and lay assessments of risk and to determine those factors that most strongly influenced public assessments of risk associated with some 90 diverse activities/technologies. Concerns about the differences in perception, meaning the various attitudes, judgements, and predispositions held by experts and by lay people, led Slovic *et al.* to develop questionnaires in order to elicit peoples' 'expressed preferences' in relation to the risks and benefits associated with various hazards and activities.[7]

Subjecting the results of a National Survey of Public Perceptions of Risk to factor analysis, the studies revealed that the following central characteristics associated with risk most strongly influenced public perceptions; these were:

- the balance of **benefits** judged to result from an activity;
- **dread**, *i.e.* the extent to which the activity of technology induced a sense of fear;
- **familiarity**, *i.e.* the extent to which the activity/technology was a part of people's everyday lives; and
- number of people exposed, *i.e.* **catastrophic potential** of the activity/technology for causing harm to large numbers of people.

Clearly, radioactive waste disposal presents an extreme picture in all four of these characteristics. Public perceptions link strongly the fear of cancer from radiation to the lack of understanding about radioactivity, and to an imagined catastrophe involving large numbers of people or a large tract of the environment. Another way of putting this is that the emphasis within public perceptions of environmental risk lies more in the area of *consequences* of the unfavourable outcome rather than upon the *likelihood or probability* of occurrence in the first place, however small. Research in this area has been extensive and a vast range of characteristics have been uncovered which are important in influencing public perceptions (see Table 2).

It has been suggested that every hazard has a particular 'personality profile'. Figure 1 represents a spacial representation of hazards within a factor space, reflecting the degree of which a hazard is understood and the degree to which it creates a feeling of dread. Research has demonstrated that public perceptions are closely related to the position of a hazard within this factor space. The following are the most important features of public perceptions of risk within this representation:

- the most influential factor is dread risk;
- the further to the right in Figure 1, the higher the perceived risk;
- the further to the right in Figure 1, the more people wish to see current levels of risk reduced and stricter regulation of the hazard imposed.

Radioactive waste disposal holds a prominent place within this factor space.

In 1983 Lindell and Earle,[9] undertook a survey to determine which factors affected people's willingness to live near a range of different industrial facilities. A

Table 2   *Factors important in risk perception and evaluation*

Factor	Conditions associated with increased public concern	Conditions associated with decreased public concern
Catastrophic potential	Fatalities and injuries grouped in time and space	Fatalities and injuries scattered and random
Familiarity	Unfamiliar	Familiar
Understanding	Mechanisms or process not understood	Mechanisms or process understood
Uncertainty	Risks scientifically unknown or uncertain	Risks known to science
Controllability (personal)	Uncontrollable	Controllable
Voluntariness of exposure	Involuntary	Voluntary
Effects on children	Children specifically at risk	Children not specifically at risk
Effects manifestation	Delayed effects	Immediate effects
Effects on future generations	Risk to future generations	No risk to future generations
Victim identity	Identifiable victims	Statistical victims
Dread	Effects dreaded	Effects not dreaded
Trust in institutions	Lack of trust in responsible institutions	Trust in responsible institutions
Media attention	Much local media attention	Little media attention
Accident history	Major and sometimes minor accidents	No major or minor accidents
Equity	Inequitable distribution of risks and benefits	Equitable distribution of risks and benefits
Benefits	Unclear benefits	Clear benefits
Reversibility	Effects irreversible	Effects reversible
Origin	Caused by human actions or failures	Caused by Acts of Nature or God

Reproduced by permission from reference 8.

central finding here was that the public *did not trust* the facility operators, so a further perceived risk factor which needs to be specified is that of personal control of consequences. This is a finding that has been supported by many subsequent studies. It also goes some way towards explaining why so many people will willingly expose themselves and their children to high levels of risk in their everyday lives, on the road, at home, and in play, if they feel they are exercising some personal choice in driving too fast, in carrying out DIY repairs in the home, and in choosing to go hang gliding or mountaineering for pleasure. They value the benefits, exercise the choice, and therefore accept the consequences.

With regard to radioactive waste disposal, where the ability to exercise personal choice is highly constrained, people's assessments of risk are strongly influenced by their perceptions of issues which they take to have an effect on them personally: for example, the perceived economic costs, say, in falling property prices near to a proposed facility; the concern for the health of identifiable persons including, old people, children; a fear of unseen hazards such as radiation and pollution of groundwater; and last but by no means least, a sense of lack of control over how the facility is located, licensed and operated. For many

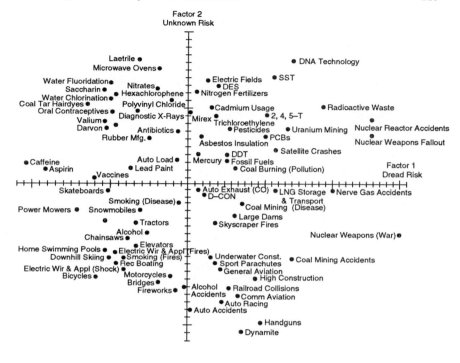

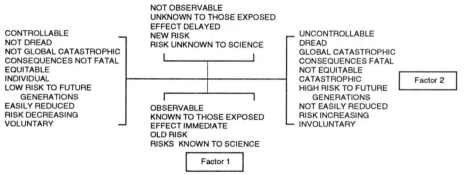

Figure 1 *Location of 81 hazards on Factors 1 and 2 derived from the interrelationships between 15 risk characteristics*
(Redrawn from reference 11, copyright 1987 by AAAS. Used by permission)
Note: Each factor is made up of a combination of characteristics, as indicated by the lower diagram.

members of the general public, these adverse consequences outweigh any perceived benefits.

The policy implications of this are significant. For instance, should investment decisions be based upon expert assessments of the likely health risks or rather upon the public's perceived assessment? McClelland *et al.*[10] provide an alternative solution, namely to invest in risk communication projects in order to attempt to reconcile residents' concerns with expert judgements about the likely impacts. The word reconcile is employed cautiously here for there is no wish to give the impression that either the expert assessment, as opposed to the public perception, is correct and that the public view needs correction; rather the point is that expert and public views of risk differ markedly and that mutual enlightenment of the alternative perspective is the essential basis for improved environmental decision-making. In the words of Paul Slovic (reference 11, p. 285):[11]

> "Perhaps the most important message from this research is that there is wisdom as well as error in public attitudes and perceptions. Lay people sometimes lack certain information about hazards, however their basic conceptualization of this risk is much richer than that of the experts and reflects legitimate concerns that are typically omitted from expert risk assessments. As a result, risk communication and risk management efforts are destined to fail unless they are structured in a two-way process. Each side, expert and public, have something valid to contribute. Each side must respect the insights and intelligence of the other".

4   Risk Communication

Risk Communication is the collective noun for a variety of procedures expressing quite different approaches towards the relationship between the general public and risk experts. An orthodox approach to risk communication, which developed in part out of the psychometric studies of risk perception, conceptualizes the risk communication process as involving an information source, a channel for transmission, a message or signal, and a receiver. An important element in this framework is the 'signal value' by which Slovic et al.[12] meant the extent to which an event or activity provides new information about the likelihood of similar or more hazardous future such events.

The signal potential of an event reflects its potential social impact and there are some interesting features here. For example, an accident that kills a large number of people may produce very little long-term social disturbance if it occurs as part of a familiar well-understood system. For example, a major air crash as a result of pilot error, or conversely a small accident in an unfamiliar or mistrusted system such as a private chemical company, may have immense social and economic consequences if it has high signal potential, i.e. if it is seen as a harbinger of worse things to come. Thus, within Slovic et al.'s factor space, Figure 1, events with high signal potential were found to predominate at the top right-hand corner of the diagram. According to Slovic:[13]

> "One implication of the single signal concept is that effort and expense beyond that indicated by a cost/benefit analysis might be warranted to reduce the possibility of 'high signal' accidents".

Thus the perspective draws attention to the types of information which, when communicated to the general public, make even wider gulfs appear between expert technical assessments and the perceptions of the general public, and of course, even political decision-takers. So this traditional approach to risk communication focuses upon carefully explaining the content of the information and getting the 'signal' right and communicated clearly to the correct 'audience'. To put this another way, the concern here is on the *content* or *what* is communicated rather than *how* it is communicated.

More recently, researchers have started to focus rather upon the *process* as opposed to the content of risk communication. The US National Research Council Committee on Risk Perception and Communication published 'Improving Risk Communication'.[14] Here, the following definitions were offered:

> "We see risk communication as an interactive process and exchange of information and opinion among individuals, groups, and institutions. We construe risk communication to be successful to the extent that it raises the level of understanding of relevant issues or actions for those involved and satisfies them that they are adequately informed within the limits of the available knowledge".

This work has been progressed further with an emerging concern on both sides of the Atlantic with questions of trust. In a recent paper, Gaskell[15] has argued that there are three basic levels of trust to consider:

Level 1 *Trust in the system*: do the basic elements of the system accord with people's values?

Level 2 *Trust in the technical competence of the experts*: do those running the system inspire confidence?

Level 3 *Trust in the public/fiduciary responsibility*: will the partners in the interaction carry out their obligations and responsibility? Will they act in the interest of the common good?

Gaskell argues that in the public domain, the decline of trust is largely at Levels 2 and 3; only the extremely alienated reject the democratic system itself (Level 1). He goes on to say that people generally accept the market system and their key concerns are actually the competence of those in charge (Level 2) and the peoples trust in a company's apparent willingness to accept their broad responsibilities (Level 3). Despite a sense of growing disillusionment and distrust, as a product of the apparent incompetence of the political system and of late a growing feeling on both sides of the Atlantic of an increasing tendency towards 'sleaze' in public life, many controversial environmental decisions are seen to be subject to political neglect rather than active political promotion. It is therefore the perceived competence of industry and of the regulators that is the concern, not only to those within the mainstream environmental groups but also to members of the public most closely affected by a particular proposal.

Gaskell[15] goes on to give advice regarding how to cultivate trust and to cultivate a sound reputation. While he saw this advice as important for private companies, they may also be pertinent to the regulator. The advice includes the following:

- Technical competence is crucial, don't deny the risks because few will believe that nuclear is risk free;
- Acknowledgement, *i.e.* admit mistakes of the past and respond to crises quickly and effectively;
- View risk management and risk communication as two sides of the same coin, they feed off each other;
- Disclosure, *i.e.* be more forthcoming about who you are and what you stand for;
- Accountability, *i.e.* articulate areas of responsibility and make sure someone at senior level is visible and accountable;
- Take on the wider responsibilities and obligations reflecting the concerns of society.

Gaskell's essential lesson is that it is important to define and understand one's public as a basis for effective communications. There are different types of supporters and opponents, different interests, concerns, backgrounds, media consumption patterns, and different levels of trust and reputation that need both history as a reliable background and history based on actions founded within a longer-term strategy of appropriate communications. Table 3 is a tongue-in-cheek look at how to *lose* trust and there are some important lessons here.

5   Radioactive Waste Disposal and Risk Communication Experience

It has been argued elsewhere (see, for example, reference 16) that in several countries the emphasis placed on independent regulatory agencies and processes has been central to the development of radioactive waste disposal proposals. The role of technical advisory groups and access to technical documentation as a means of creating public trust in the technical competency of the proposals and of their scrutiny, does vary, however, from country to country. The use of advisory panels, citizens groups, local liaison committees, site-selection commissions, public hearings, and negotiations with host communities also varies. So there is no universally accepted model which draws such mechanisms together in a uniform or coherent way. There is no universal panacea to the question of risk communication and public acceptability for radioactive waste disposal.

However, it has been possible to delineate a range of decision-making processes through which different countries cover various approaches. At the extremes, this range is generally between devolved or centralized and open or closed decision-making procedures. At the devolved end of the spectrum the emphasis is on early public access to the decision-making process and to the regulators, and on public consultation with an incremental approach towards the final choice. I have termed this the 'establish criteria, consult, filter, decide'

Table 3   *Ten ways to lose trust and credibility*

Take a good look at most risk communicative 'horror stories' and you'll probably find a major breakdown in trust between government representatives and the public they are supposed to serve. The next time someone comes to you with a sob story about communicating with the public, you might want to hand them this tongue-in-cheek list. Or better yet, hand it out before the damage is done.

1. **Don't involve people in decisions that directly affect their lives.** Then act defensive when your policies are challenged.
2. **Hold onto information until people are screaming for it.** While they are waiting, don't tell them when they will get it. Just say, "These things take time," or "it's going through quality assurance".
3. **Ignore people's feelings.** Better yet, say they are irrelevant and irrational. It helps to add that you can't understand why they are overreacting to such a small risk.
4. **Don't follow up.** Place returning phone calls from citizens at the bottom of your 'to do' list. Delay sending out the information you promised people at the public meeting.
5. **If you make a mistake, deny it.** Never admit you were wrong.
6. **If you don't know the answers, fake it.** Never say "I don't know".
7. **Don't speak plain English.** When explaining technical information, use professional jargon. Or simplify so completely that you leave out important information. Better yet, throw up your hands and say, "You people could not possibly understand this stuff".
8. **Present yourself like a bureaucrat.** Wear a three-piece suit to a town meeting and sit up on stage with seven of your colleagues who are dressed similarly.
9. **Delay talking to other agencies involved,** or other people involved within your agency, so the message the public gets can be as confusing as possible.
10. **If one of your scientists has trouble relating to people, hates to do it, and has begged not to, send him or her out anyway.** It's good experience.

approach,[16] which is the essential process that has been followed in Canada and Sweden. At the centralized end of the spectrum the emphasis in on minimizing public involvement and resting authority and control firmly in the hands of the responsible organizations. Public acceptance is sought through firm leadership. This approach I have termed the 'decide, announce, defend' system and is more characteristic of procedures followed in the USA and the UK. However, it is clear that most countries lie somewhere within this spectrum and not necessarily at the extremes.

Clearly, there is a wide range of potential processes between the two extremes. I have argued elsewhere[16] that the key consideration is not necessarily where a particular approach can be placed on the spectrum, but in *which direction* that approach is moving. This directional commitment relates to the ranges of mechanism employed for generating public acceptance. In addition, the consistency with which a particular direction is maintained also bears upon public acceptance. Vacillation between open and closed approaches results in the undermining of public trust in the regulatory authorities. It is here that the UK approach requires careful attention.

The Approach to Probabilistic Assessment for Radioactive Waste: The UK Philosophy

In a recent editorial, Bonano[17] and Thompson have provided an essential insight into current thinking about the requirements for carrying out the full long-term radiological performance assessment for deep disposal. The authors draw attention to several elements, which they argue should all be included. They note a regulatory requirement in terms of constraints and decision criteria, and the need for a comprehensive representation of the repository, its wastes and its evolving surface and sub-surface environmental setting as a mathematical model of the entire system. They point to the necessary computational tools and techniques that should be designed, developed, tested, released, documented, and applied in a manner that ensures confidence and traceability. A procedure for performing the assessment, which can account explicitly for the large-scale uncertainty and subjectivity that is unavoidable in any long-term forecasting problems, is recommended. Similarly, they stress the need for a framework for relating the performance assessment as such to the site investigation programme, and to the supporting research and development activity, in order to ensure maximum efficiency in reducing overall ignorance and uncertainty, which affects the decision process.

More pertinently, this approach has a number of interesting resonances with the arguments laid out above on risk perception, communication, and public trust. In particular, the authors emphasize the need for:

- a means whereby independent peer review of the important subjective judgements underlining the analysis is ensured;
- communication during the overall regulatory process with different stakeholders, each with their different perception of risk, so as to fold in social, psychological, and economic influences (reference 17, p. 103).[17]

It is interesting to note the emphasis on the need for comprehensiveness, traceability, reducing uncertainty, independent peer review, and on-going communication. These are the key principles upon which the commentators and authors from a range of different scientific and social scientific perspectives are now reaching consensus. The challenge is to devise the practical means for this implementation and achievement, especially in relation to the final two points highlighted above.

6   Conclusions

To the extent that consensus is emerging on the fundamental requirements of an acceptable risk communication process for problematic environmental risk issues, there now remain a number of practical questions to be addressed. While it will fall to others to lay out more clearly the various dimensions of this debate introduced above, it is the author's view that essential areas for discussion include the following.

- *What can be learnt* from the four paradigms of risk in communicating the results of a risk assessment? In the first paradigm, quantified risk assessment generates risk numbers as a basis for judgement. But acceptable or tolerable risk depends upon:
  (i) *how* the number is obtained/derived; and
  (ii) other social, psychological, cultural, and organizational factors as suggested by the subsequent paradigms, which are not readily expressed in numerical form.
- *Understanding the language or the vocabulary of risk.* How do various groups understand terms such as 'risk', 'probability', 'conservatism', 'uncertainty', 'failure', and so on requires attention?
- *Explaining concepts* of quantified risk. What are the most effective forms of explanation?
- *Dealing with concepts of time and space in risk.* What is the ability of different groups to conceptualize geologic time and three-dimensional space? Alternative means of presenting the results of risk assessments via visual simulations, computer graphics, simple documentation, simple verbal presentations, *etc.* must be explored.
- *Appropriateness of documentation/reports etc.* for explanation/communication of the content and competency of risk assessments in *different arenas* for different audiences;
- *Assessing the usefulness of different technical working arenas* for the discussion and presentation of the results of risk assessments, for instance working seminars, round table discussions;
- *Anticipating the formal requirements of the public hearings process.* The British public local inquiry system can accommodate a range of different types of session, ranging from the formal presentation of evidence under cross-examination, through to side-room meetings between experts to identify areas of agreement and disagreement; what is the most effective means?
- *Preparing for alternative challenges* such as the likelihood of an alternative 'unofficial' public inquiry being held;
- *Building trust* through association with other trustworthy sources and parties such as the role of independent or highly respected expert groups, and the use of locally recognized individuals and/or respected institutions;
- *Demonstrating accessibility* and openness in the role of the regulatory authorities and other responsible parties;
- *Consideration of alternative mechanisms* for communicating with various groups of the public such as the use of focus groups, planning cells, or other non-traditional approaches to communicating with the local community.

References

1 T. Kuhn, 'The Structure of Scientific Revolutions', University of Chicago Press, Chicago, IL, 1962.
2 'Risk Assessment', Report of a Royal Society Study Group, The Royal Society,

London, 1983, p. 157.

3 T. O'Riordan, R. Kemp, and M. Purdue, 'Sizewell B: Anatomy of the Inquiry', Macmillan, London, 1988.

4 Health and Safety Executive, 'The Tolerability of Risk from Nuclear Power Stations', HMSO, London, 1988.

5 R. E. Kasperson, O. Renn, P. Slovic, et al., 'The social amplification of risk: a conceptual framework', Risk Anal., 1988, 8, 177–187.

6 B. Fischoff, S. Lichenstein, P. Slovic, S. L. Derby, and R. L. Keeney, 'Acceptable Risk', Cambridge University Press, Cambridge, UK, 1978.

7 P. Slovic, B. Fischhoff, and S. Lichtenstein, 'Regulation of risk. A psychological perspective', in 'Regulatory Policy in the Social Sciences', University of California Press, Berkeley, 1985.

8 V. T. Covello, P. M. Sandman, and P. Slovic, 'Risk Communication, Risk Statistics and Risk Comparisons; A Manual for Plant Managers', Chemical Manufacturers Association, Washington, DC, 1988.

9 M. K. Lindell and T. C. Earle, 'How close is close enough: public perceptions of the risks of industrial facilities', Risk Anal., 1983, 4, 235–253.

10 G. H. McClelland, W. D. Schulz, and B. Hurd, 'The effects of risk beliefs on property values: a case study of a hazardous waste site', Risk Anal., 1990, 10, 485–497.

11 P. Slovic, 'Perception of risk', Science, 1987, 236, 280–285.

12 P. Slovic, S. Lichtenstein, and B. Fischhoff, 'Regulation of Risk. A Psychological Perspective', in 'Regulatory Policy in the Social Sciences', ed. R.G. Noll, University of California Press, Berkeley, CA, 1985.

13 P. Slovic, 'Perceptions of Risk: . . .', in 'Theories of Risk', ed. D. Golding and S. Krimsky, Auburn House, Dover, MA, 1991, p. 120.

14 US National Research Council Committee on Risk Perception and Communication, 'Improving Risk Communication', US National Research Council, 1989.

15 G. Gaskell, 'In Safe Hands: Building the Public's Trust', Paper to 'Opinion '94: Attitudes to Nuclear Energy', British Nuclear Forum, House of Parliament, Westminster, London, 1994.

16 R. Kemp, 'The Politics of Radioactive Waste Disposal', Manchester University Press, Manchester, UK, 1992.

17 E. J. Bonano and B. G. J. Thompson, 'Probabilistic risk assessment for radioactive waste', Reliability Eng. System Safety, 1993, 42, 103–109.

Chapter 16

Risk Assessment and Reality: Recognizing the Limitations

By Paul Johnston, David Santillo, and Ruth Stringer

GREENPEACE RESEARCH LABORATORIES, UNIVERSITY OF
EXETER, NORTH PARK ROAD, EXETER, EX4 4QE, UK

1 Introduction

Current approaches to pollution control and environmental protection are underpinned by attempts to manage chemicals that enter the environment and to thereby mitigate potential impacts upon ecosystems and human health. At least two frameworks operate in the European Community (EC). One is based upon unified emission standards (UESs) from point sources and is applied by most European regulatory authorities. The second is applied by the UK authorities and is based upon a system of environmental quality objectives (EQOs) and standards (EQSs) for specific chemicals.[1,2] The ultimate object of both approaches is the control of chemicals. The UES approach seeks to regulate at the point of discharge while an EQO/EQS approach regulates levels of chemicals in the wider environment. In practice, however, neither approach can be adopted in isolation. A fixed emission framework still requires monitoring in the wider environment to be carried out while environmental quality standards can only be met through enforcement of conditions imposed upon point source discharges of chemicals. In some cases, such as control of microbiological contaminants for protection of fishery and amenity resources, environmental quality standards are applied throughout the EC.[3]

The selection of chemical contaminants requiring priority control has largely been based upon whether or not they exhibit the combined properties of toxicity, environmental persistence, and a potential to bioaccumulate. Other commonly used defining properties in animal and human models are carcinogenicity, mutagenicity, and teratogenicity. Some form of regulatory prioritization has been inevitable given that most regulatory instruments are permissive in nature and formulated to allow chemicals to be released into the environment. It is estimated that some 63 000 chemicals are in common use worldwide, that about 3000 account for 90% of the total production, and that anywhere between 200 and 1000 new synthetic chemicals enter the market each year.[4] Other figures suggest that in the EC alone, 50 000 substances are in use of which 4500 are demonstrably toxic, persistent in the environment, and may be bioaccumulative.[5] Prioritization procedures depend upon evaluation of a chemical in relation to defined values of these parameters. One well-known example is the UK 'Red List' system[6] where numerical thresholds for each factor are used in order to establish the necessity of priority regulation and control.

Existing frameworks for chemical regulation and control are, therefore, essentially based upon hazard assessment. This methodology for analysing the environmental effects of chemicals depends upon a comparison of expected concentrations of the chemicals in the environment with the concentrations (usually estimated) at which toxic or other defined environmental effects can be observed. These are not probabilistic techniques: they depend upon simple analyses of data supplemented by expert judgment, application of 'safety factors' to accommodate uncertainty, and a commitment to iterative refinement of the data.[7] In many cases, surrogate values are used to estimate some of these properties, for example predictions based upon chemical structure, known as quantitative structure–activity relationships.[8] On this basis, hazard becomes a property of a given chemical, defined in terms of the parameters selected for its evaluation. The process of assigning hazard ratings can be extremely slow. From the priority candidate list of 129 chemicals selected by the EC from an original subgroup of 500, only 18 have been classified as EC List I chemicals under EC Directive 76/464/EEC.[5]

Risk assessment has been described[7] as an array of methodologies that has developed from actuarial techniques used by the insurance industry. In essence, it holds as a central thesis that it is possible to objectively define the uncertainty attached to undesirable changes taking place in the environment as a result of anthropogenic actions. Simply, it attempts to define the probability of undesirable effects occurring in the environment. This has largely impacted upon the issues concerning human health, where particular health outcomes resulting from a given environmental exposure are reduced to a probabilistic function. This is most often seen in the form of a stated risk of contracting cancer. A one in a million risk is most often used as the delineating probability between acceptable versus unacceptable risk under these circumstances for the US population[9] although risks of some adverse outcomes as high as 10^{-2} and 10^{-3} have been allowed for relatively small, exposed populations. In estimating risk, as opposed to hazard, it is necessary to have information concerning exposure of the studied system to a given chemical.[9]

The rationale that addressing human health concerns will of itself confer protection upon the wider environment has long been suspect as a premise. Indeed the only area where this philosophy overtly persists is in the regulation of radionuclides.[10] In other fields of environmental research it has been recognized that other factors need to be taken into account that may not be immediately obvious from considering the human health aspects alone. These include the greater sensitivity of organisms other than humans to a given exposure, the lack of ecosystem analogues of many human responses, and the converse situation, potentially more intense ecosystem exposures and the greater degree to which non-human organisms are coupled to their environment.[7] Hence, there is a growing impetus to apply risk assessment techniques to the evaluation of risks to the wider environment and this is generally referred to as ecological risk assessment. In its purest, intended, form this is regarded as a rigorous, scientifically robust, numerical technique that explicitly requires a robust evaluation of exposure as a key element.

There are considerable attractions to the use of risk assessment for regulatory agencies and regulated industries alike. Their use is justified as a way of simplifying and expediting regulatory effort and introducing a quantified predictive element into the equation. In addition, both consider that risk assessment can be used to accommodate uncertainties in the scientific data and that risks can be assigned defined ranges by identifying and manipulating variables whose precise characteristics are uncertain. The techniques, inevitably, have been evolved to accommodate economic variables and this is described as risk–benefit analysis. Management techniques based upon risk–benefit evaluations seek to identify an optimal level of environmental protection. This is when any risks have been reduced to the point where the costs of any further reduction just equal the benefits involved. A risk–benefit analysis cannot ascribe a fiscal value to risk. Where this can be done, however, a cost–benefit analysis results.[11] Hence, risk–benefit analysis is regarded as a component of cost–benefit assessment.

This departure of the techniques of risk assessment from a purely probabilistic calculation into the realms of the commercial and fiscal has raised a number of concerns, particularly since the cost–benefit concept is now incorporated into the definition of the precautionary principle[12] as defined by the UK Government. It can be argued that such profound alterations to the underlying philosophy will force environmental quality towards a series of lowest common denominators rather than promote wholesale improvement. Further, they raise the contentious issue of assigning notional values to both environmental quality and human life. In short, they blur the distinction between the purely numerical evaluation itself and the use of the derived numerical value to formulate policy. While these issues are of considerable societal importance, it is not intended to examine them in any greater detail here. Arguably, however, such contentions might have less significance if the underlying assessment of actual risks in quantitative terms was plausible. In practice this critical element is demonstrably flawed in a number of respects.

This chapter, therefore, outlines the considerable difficulties in attempting to quantitatively define environmental risks arising from chemicals in the environment. These may be conveniently divided into the following categories: analytical and monitoring problems, problems with evaluating the effects of chemical mixtures, and problems in defining appropriate end-points of response to exposure.

2 Analysis and Monitoring

In estimating the potential effects of chemicals using a risk assessment framework, recourse will be made to analytical chemistry techniques. A typical sequence might be to determine the concentration of a chemical in marine food resources and then subsequently to use this to calculate population exposure via the food chain.[9] In other situations, exposure through air, water, or dermal pathways may be considered. In each case, establishing a plausible exposure value depends upon the quality of the analysis carried out.

The problems in this area may be illustrated by the difficulties encountered in protection of the North Sea over the last decade. These have been brought to the fore by the studies carried out under the auspices of the North Sea Task Force.[13,14] The major control on chemicals entering the system depended upon reducing inputs of selected contaminants by a percentage ranging between 50 and 70%. Various governments supplied data suggesting that these input reduction targets had been met. Scrutiny of the data available upon which to base any evaluation of whether this was the case, however, revealed a number of difficulties. These relate both to the accuracy of the estimates of chemical inputs and to determination of levels of residues in various environmental compartments. This in turn suggested generic limitations which undoubtedly apply to other regional pollution control initiatives and to risk assessment methodologies.

Input Estimates

In theory, input estimates for chemicals may be based upon analysis of point source discharges, or upon measurements taken in the receiving environment. A point source needs to be monitored at least 60 times in a year in order to reliably ensure statistical detection of a fall from 95% compliance on an annual basis. Below this figure, the certainty of ensuring compliance with licence conditions falls off rapidly.[15] Very few effluents are monitored so intensively. Moreover, the licence conditions themselves often regulate only a small proportion of the chemicals routinely discharged.[16]

According to the UK Royal Commission on Environmental Pollution, even licences regulating large industrial discharges may routinely control fewer than 20 physico-chemical determinands.[17] In addition the Commission report also notes the difficulty of evaluating diffuse sources of contaminants. Data generated by the more widely based monitoring programmes do not provide reliable information on inputs. The intensity and frequency of monitoring of even the limited number of chemicals contained in the UK 'Red List'[6] is not regarded as sufficient to detect reductions of inputs of chemicals on a local basis, let alone over the North Sea as a whole.[18,19] As another example, in the case of the Tyne and Tees rivers, metal loadings have been computed for input evaluation on the basis of just four samples a year.[20]

Unsurprisingly, the published input estimates of chemicals into the North Sea are subject to wide uncertainty. In some cases, the maximum and minimum input estimates differ by a factor of 10. In the 1987 quality status report it is noted that all input figures with the possible exception of dumping are subject to considerable uncertainty, the extent of which is variable and difficult to quantify.[13] This has been apparent since at least 1984 when the first estimates of contaminant inputs from the UK became available.[21] There appears to have been no improvement in this situation to date. The reliability of loading figures is still low and recorded changes in annual loads from rivers are seldom of statistical significance.[20] Certainly, therefore, little justification exists for claiming that, for example, cadmium in the North Sea has reached a steady state.[22]

Atmospheric inputs comprise a further confounding factor. One of the more important results of the recently reported 5 year UK Natural Environment Research Council Project was the possibility that in excess of 50% of some metals entering the North Sea do so from the atmosphere.[23] It seems likely, therefore, that lack of plausible input data exists for the majority of receiving waters on a global basis since the North Sea environment is comparatively intensively monitored. Under these circumstances, protective instruments based upon measured percentage reductions are compromised, while any risk assessment procedure dependent upon such data is inevitably fatally confounded. Much lauded percentage reductions in chemical input, moreover, may well be a complete artefact of the data.

Quality of Analytical Data

In turn, the quality of the data derived from analytical programmes needs to be viewed with a great deal of suspicion. The project manager of the latest international Quality Control and Quality Assurance Project (QUASIMEME) notes that it has become quite clear that the measurements made on the same sample by one laboratory often bear no resemblance to the values obtained by another laboratory. More worrying still, this means that the interpretation of the environmental significance of such data sets is no more than an observation on the spread of errors generated by the scientists concerned.[24,25] This is clearly illustrated also by data gathered in a German intercalibration exercise for individual polychlorinated biphenyl (PCB) congeners.[26] The data show a wide range and extreme erroneous values were generated, but not consistently, by any single laboratory. The report concludes that any absolute comparison of the data obtained is only justified if statistical variability is accounted for and that the same PCB congeners are compared. These problems are common both to the North Sea Task Force and to the Joint Monitoring Group of the Oslo and Paris Commissions active in the North East Atlantic region. In 1992 this last group reported a great improvement in the mutual comparability of analysis results. The standard deviations for most PCBs in seal-blubber and sediment extracts had diminished to around 20%.[27] It follows that even if data do stretch back a number of years[28] there has as yet been no comprehensive intercalibration exercise carried out to validate them. Therefore, the use of historical data in resolving trends in contaminant levels and hence in estimating exposures is limited.

Similar problems emerge when considering data sets obtained from long-term monitoring of contaminants present in biota. In this case the analytical imprecision is compounded by others resulting from biological variability. Difficulties in detecting trends are identified in a recent assessment of the ICES Cooperative Monitoring Programme (CMP) of contaminant levels in fish muscle.[29] It is stated that, accounting for random between-year variation, it would be unrealistic to expect to detect any patterns of change over a 2 to 3 year period. The CMP data sets span no more than 8 years which would give sufficient power to the programme to detect trends in zinc of 10% per year and of copper and mercury of 20% per year, assuming that analytical data are reliable. These are large trends and this has significant implications for exposure estimates. If large-scale pollution incidents are considered, then zinc and copper would need to change by 200% and mercury by 400% to be identified. Only by extending the CMP data to cover 20 years does the power of the programme rise to the extent that trends of 5% can be detected. As pointed out by the authors, the environmental impact of a contaminant over this period, prior to trends emerging, may well prove unacceptable.

Recent serious complications have emerged also in the use of fish as 'sentinel species' for pollution. It has recently been found that for the target species of a US benthic surveillance programme the length of the fish alone accounted for between 19 and 67% of the variability in the data. The study considered only metals and was initiated in response to concerns that size differences alone between fish sampled in different years could obscure the identification of spatial differences or temporal trends.[30] A report from the Netherlands noted that levels of DDT [1,1,1-trichloro-2,2-bis(*p*-chlorophenyl)ethane] and PCBs in cod liver were found to be related to the length or age of the sampled fish. A recalibration of the trends data was necessary since average lengths of the fish have decreased considerably during the last decade.[27]

Implications for Exposure Estimates

Overall, there appears to be little hope of characterizing exposure of organisms or trends in exposure on an ecosystem-wide basis with any great certainty. Under these circumstances, a degree of extreme subjectivity exists in any risk assessment. It is difficult to see how this problem can be fully rectified. Intensification of wide field monitoring regimes is likely to be a cost-intensive procedure. Cost penalties will also accrue from more intense oversight of point source discharges. In the latter case, however, improvement in data resolution is likely to increase to a greater extent for a given unit cost.[31] This is simply because analysis of concentrated discharges poses less technical difficulty than analysis involving samples taken from open waters after considerable dilution has taken place. However, it can be argued that accurate data on contaminants in open waters gives a more reliable indicator of ecosystem exposure. In essence, point source monitoring implies a bias towards control of chemicals, while the monitoring of ecosystem levels draws environmental protection further towards a strategy of management of chemicals in the ecosystem.

It is often supposed that improvement and verification of analytical methods will, in time, partially solve these problems. Nonetheless, there are likely to be surprises. A good example of what can happen is provided by the results of the Total Exposure Assessment Methodology studies conducted by the US

Environmental Protection Agency (EPA).[32] These became possible after the development of suitable personal air samplers. The studies were conducted with two basic guiding principles that firstly all exposure should be measured directly wherever possible and secondly that all participants were to be selected according to robust statistical protocols. The studies focused on around 20 volatile organic compounds. Some long-standing assumptions concerning exposure to these chemicals were overturned. Although the petrochemical industry releases the greatest quantities of benzene, exposures were largely attributable to active and passive smoking. Similarly, the major environmental releases of perchloroethylene are from industry and dry cleaning establishments, yet population exposure is largely attributable to the chemical being introduced into the home on dry-cleaned clothes. Exposures to *p*-dichlorobenzene from domestic products were greatest in the home, as were exposures to certain chlorinated pesticides some of which, like DDT, had been banned a long time previously.

These intriguing results led to the conclusion that exposures must be measured directly, that the amount of a pollutant released by given sources may be subject to considerably different 'delivery efficiencies', and that sources of chemical contaminants in the home must be considered in any exposure assessment and subsequent risk evaluation. In addition, it raises the question of whether personal exposures are adequately controlled by regulation of manufacturing sites or whether the use of certain chemicals in domestic products needs much tighter control. It is likely to be far more difficult to evaluate exposures of organisms other than man in various ecosystems on this basis since such intensive sampling is rarely possible and pollutant dynamics are more complex and less easily evaluated. Overall, of course, a failure to derive credible exposure data means that a risk evaluation cannot be developed beyond a putative hazard assessment.

3 End-points and Markers of Effect

Conceivably, it could be argued that uncertainties in exposure data could be compensated for. Exposure could be manipulated as a variable within the risk assessment framework. This, however, is a departure from the central paradigm of toxicology that dose (exposure) determines effect. In addition, such manipulation actually violates the critical commitment in risk assessment to establishing empirical exposure values.[9,11] An alternative approach is to look for both generalized and specific end-points of response to toxicological stress and monitor these in the environment. In cases where there is a specific response, such as the formation of DNA (deoxyribonucleic acid) adducts upon exposure to benzo[*a*]pyrene or the presence of imposex in a population of marine molluscs exposed to tributyltin this may be a useful approach. This is only useful, however, where causality can be unequivocally assigned, *i.e.* where specific effects result from specific exposures. In practice, such direct relationships are rare, even in workforces occupationally exposed to chemicals, and certainly in natural ecosystems.

The search for suitable end-points in the toxicity test procedures applied in hazard assessment and subject to iterative verification has provoked much discussion in the literature.[33] Unsurprisingly, this discussion has extended into ecological risk assessment. An excellent overview of the subject of ecotoxicological end-points has recently been published.[34] This distinguishes the substantial differences between end-points derived from toxicity tests [such as LC_{50} (dose that is lethal in 50% of test subjects) values] and assessment end-points. Assessment end-points nearly always refer to effects upon higher levels of ecosystemic organization and over larger spatial and time frames in comparison with the most sophisticated simulated systems used in tests.

The point is made that ecosystem interactions are both subtle and complex. Extrapolation from test end-points to predict ecosystem effects may be valid under certain limited circumstances. Generally, however, constraints exist that limit the predictive power of test end-points. These constraints, or the way in which they operate, may differ markedly in a simulated ecosystem with respect to real-world systems.

The end-points themselves fall into four broad categories. Sub-organismal end-points involve biochemical, physiological, or histological parameters, referred to as biomarkers. Although important, they are not usually of interest to environmental managers in themselves, and moreover there are few predictive models capable of using such data as input. Organismal assessment end-points, which measure death, lifespan, reproductive vigour, or behavioural responses, are most often inferred from standard toxicity test data and cannot, in the vast majority of cases, predict ecosystem effects. Population responses are favoured by environmental managers but toxicity tests are rarely designed to generate population level data. These parameters are relatively chemically non-specific. Nonetheless population level responses are frequently used to assess responses at the next level of organization: whole ecosystems. Assessment at this level is fraught with difficulties since both assessment and test end-points for these systems are problematical and ill-defined. Hence, a general strategy is to select ecosystem end-points on the basis of management goals. Inevitably, it is observed that use of ecosystem end-points in assessment of ecosystems has been simplistic. There is no realistic way of assigning causality by using what can be described as ecological epidemiology.

With each layer of ecosystem organization, therefore, there is a progressive move away from end-points which have utility in the assignment of causality. The problems are equally apparent in human epidemiology.[35] The difficulties of establishing causality in human populations mean that there is a high degree of uncertainty about these relationships. There are relatively few well-defined end-points, and a high dependence upon carcinogenic effect as an index.

An interesting recent natural ecosystem study demonstrates how the sensitivity of an assessment may be influenced by use of more sensitive end-points and more sensitive statistical analysis. Organisms living in the vicinity of oil and gas production platforms in the North Sea are impacted by drilling muds and operational discharges. Early studies suggested that normal baseline biological conditions were attained within 200–1000 m of the 500 m radius primary impact zone. By using more sensitive biological indices and multivariate statistical techniques, the area of biological impact extends up to at least 3 km from installations.[36] This estimate has been further refined upwards to encompass a 6 km radius around some installations in an elegant demonstration of the fact that assessment of impact upon which future risk assessments depend is highly dependent upon the techniques used to investigate it.[37]

Endocrine Disrupters

Recently, a number of reports have appeared that outline a phenomenon with highly significant implications to natural ecosystems and humans alike.[38–40] It has become apparent that some chemicals released into the environment have the ability to disrupt endocrine systems. This disruption may affect a wide range of organismal functions but has been most readily observed in the form of reproductive disturbance in fish exposed to a range of chemicals in sewage and industrial effluents. Some of the chemicals potentially responsible have been identified. These include various organochlorine pesticides, the PCBs, the dioxins, the phthalate plasticizers, and some phenolic compounds. It is clear that others remain to be identified. Concerns have been voiced that such chemicals may be responsible for observed declines in male fertility in humans and contributors to the aetiology of cancers and reproductive disease.

Endocrine disrupters illustrate the inherent paradoxes in conventional risk assessment procedures. They induce an end-point at the sub-organismal level by interfering with hormonal pathways. Organismal responses have been identified that may affect whole populations. The effect upon ecosystems is largely unknown, although many speculative negative impacts can be inferred. These effects may take place at very low concentrations. The dose–response relationship is unclear and hence the classical assumptions underpinning toxicity are undermined. Exposures, therefore, remain uncharacterized and not amenable to assessment. However, a wide variety of chemicals ranging from herbicides to pharmaceutical preparations are specifically designed to interact with hormonal systems. Until recently, however, the unintended wider spectrum end-points at the sub-organismal level were completely unknown. Most importantly, there is no way that the potential for endocrine disruption can be predicted upon the basis of chemical structure alone. This means in turn that surrogate predictive tools cannot be used. In terms of actually identifying the chemicals responsible, it means that assessment frameworks come into conflict with the issue of chemical diversity.

4 Chemical Diversity

A highly conspicuous feature of toxicological data is that it largely relates to the effects of single chemicals. Very few data have been derived from the testing of combinations of chemicals. Chemicals do not occur singly in the environment and this creates yet further uncertainty in risk assessment procedures. The precision with which causality may be assigned decreases as the level of ecosystem end-point increases. Equally, precision also decreases when chemicals are present as complex mixtures.

This situation is intensified by the relatively few chemicals monitored in the environment on a routine basis and intercalibrated basis. The QUASIMEME programme, for example, embraces only a limited subset of priority chemicals. As mandatory determinands, the metals Cd, Cu, Hg, Pb, Zn are covered, together with four chlorinated species. Laboratories participating can select voluntarily from a list of 12 other chemical types. Significantly, this voluntary list includes the dioxins, which were scheduled for a 70% reduction under the terms of the 1990 North Sea Ministerial Agreement.[41] This is an extremely limited subset of the number of chemicals routinely discharged into the marine environment. Obviously, a wider variety of other chemicals are monitored in the environment, but not on a systematic basis that allows whole ecosystem impact to be assessed. Indeed in most cases these chemicals are regulated on the basis of surrogate parameters such as those defined in the UK Red List, and this has not allowed evaluation beyond the sub-organismal or organismal levels.

Analysis of effluents from a wide variety of industrial sectors has shown that a substantial proportion of compounds present are difficult to identify even using sophisticated analytical techniques. There are a number of reasons for this, but if it is not possible to identify a chemical being discharged, then it is impossible to ascertain what risk it imposes. It is highly unlikely that even effects at the lowest levels of ecosystem organization can be assessed. In addition, there are often many more chemicals present than are regulated by the licence.[16] This point is illustrated by Figure 1 which consists of three traces obtained by analysis of the solvent extracts of three environmental samples. Table 1 shows that only a relatively small proportion can be identified with a high level of certainty using GC–MS techniques.

The occurrence of complex mixtures of industrial chemicals in open waters was confirmed by research carried out by the UK Ministry of Agriculture Fisheries and Food.[42] Not only did this study identify a wide range of chemical contaminants of clear industrial origin in offshore waters, but also it showed that the exact combination varied between estuaries, presumably reflecting differing industrial activities within each area. A similar finding has been made for the Scheldt estuary. This is regarded as heavily polluted, and studies of organic micropollutants have revealed that a significant component of the chemicals isolated could not be identified.[43] A wide range of organic micropollutants have also been identified in open waters off the Dutch coast.

Toxicological Effects of Complex Mixtures

Attempts to assess the risks posed by mixtures of chemicals have been confounded by the absence of information on the properties of the individual components. It was found that for 75% of the chemicals isolated and tentatively identified in the Tees area by the sophisticated GC–MS techniques used, no ecotoxicological data could be found.[42] In many cases, due to the lack of ecotoxicological information

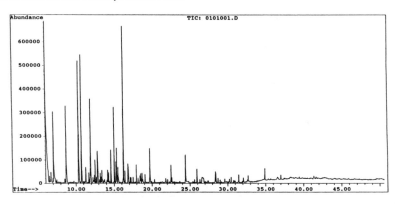

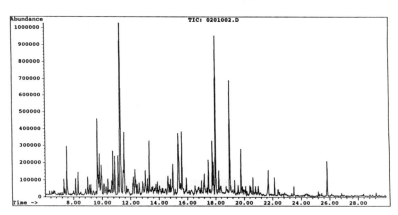

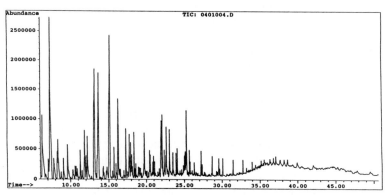

Figure 1 *Traces derived from the GC–MS analyses of solvent extracts of environmental samples. Methods are as outlined in reference 16. The traces were derived (from top to bottom) from a sewage treatment works in the UK, a freshwater sediment from an industrialized area in Israel, and a sediment obtained from a waste water evaporation pond in Spain. Each peak on the trace theoretically represents a single chemical compound. It is clear that there is considerable variability between the effluents with respect to their constituent components*

Table 1   *Numbers of chemicals identified from the traces shown in Figure 1.*
Mass spectra for each of the peaks were compared with the 136 000
specimen mass spectra held on the Wiley Spectral Library using
probability based matching techniques. A match at 90% probability is
regarded as a reliable. At 50–90%, identification is only tentative while
below 50% the compound is regarded as unidentified. Relatively few
chemicals of those present are routinely identified.

Sample	Compounds isolated	Match at 90%	Match at 50–90%	Match below 50%
Sewage	77	23	15	39
Sediment	55	15	7	33
Sediment	180	33	21	126

the toxicity of many chemicals had to be inferred by using figures derived from theoretical toxicity models based upon chemical structure.[44] The concentrations of the individual chemicals were not thought to be of concern. Subsequently attempts were made to investigate organismal responses to the mixture using sensitive toxicity tests. These were carried out using oyster embryos placed in the Tees estuary itself. Significant mortalities of the test species took place in the chemical mixture. Indeed, at one point in the upper Tees estuary in 1990, 100% mortality of the test species took place. The study, understandably, concluded by noting that accurate prediction of the joint effects of complex mixtures of substances is not possible at present.

The problems of relating chemical contaminant effects in wider natural communities of organisms are discussed in depth in a recent review.[45] These workers consider that there is a need to determine 'critical body residues' of contaminants in a comprehensive manner as an additional component of ecotoxicological assessments. They suggest that this would greatly assist the evaluation of mixture toxicity. Nonetheless, as pointed out earlier in a submission to the 1990 Ministerial Conference,[46] the lack of information about the environmental behaviour of chemical mixtures in association with other environmental stressors is a continuing and critical deficiency of evaluation procedures applied to marine and other environments.

The acquisition of further information to input into risk assessment processes may be of limited value without a wholesale effort to elucidate toxicodynamics. Mixtures of chemicals in some environmental matrices may exhibit marked departures from behaviour predicted under the assumptions inherent in toxicity tests. Both intuitively and empirically based understandings of the role of dilution may not be valid, a somewhat serious complication. For example, a recent study[47] of the immune capability of shrimp exposed to highly contaminated sediments from Rotterdam Harbour has demonstrated this quite clearly. When exposed to 100% dredge spoil the immune system of the organisms was compromised in relation to control animals living on clean sand. Unpredictably and very surprisingly, shrimp exposed to a mixture of 95% clean sand and only 5% dredge spoil showed a level of immune system compromise greater than those exposed to the 100% dredge spoil. The reasons for this are not clear but neither is it an isolated observation. This phenomenon has considerable implications for risk assessments carried out in relation to the dumping of contaminated dredge materials at sea.

5 Overview

The foregoing discussion appears at first sight to be unremittingly negative in its view of the sciences of toxicology and ecotoxicology. This is in many ways unfair. Toxicological research and attempts to evaluate ecosystem response have contributed much to our understanding of the effects of polluting chemicals. Indeed such studies, however imperfect, are our only basis for evaluating the effects of chemicals in the environment. Providing that the limitations are understood and respected, then the results from such studies may serve as useful indices in certain applications and contribute much to environmental protection.

Currently, however, risk assessment is not an application that can be considered to be well served by toxicological and ecotoxicological data. As discussed above, inadequacies exist in the data for all but a limited subset of those chemicals emitted to the environment. Exposure estimates are confounded by imprecise estimates of the input of chemicals to ecosystems and lack of direct measurements. The estimation of residue trends is subject to similar imprecision and consequently to an identified and well-recognized utilitarian uncertainty. The end-points targeted by testing procedures do not readily correspond to end-points required by assessment procedures. Assessment frameworks cannot accommodate operative functions that have not been identified, or newly recognized end-points such as endocrine disruption. Existing frameworks cannot assess the behaviour of a single chemical present in a mixture or indeed that of the mixture itself. Finally, risk assessment cannot be applied to chemicals that cannot be readily identified.

A daunting array of inadequacies and uncertainties exist, therefore, each of which on its own could compromise a risk assessment procedure. To put this in perspective, Figure 2 reproduces a conceptual risk assessment structure for human health risks from contaminated marine resources.[9] Considering the potential problems in combination, it is likely that many risk assessment procedures of this kind will be fatally compromised, but with the added danger that the compromises will not be recognized. This may lead in turn to inappropriate environmental management decisions. Even where uncertainties are recognized and can be manipulated as variables within the assessment framework to produce a range of values for risk factors, there is the attendant possibility that manipulation of more than one factor will produce equivocal results. This is not reassuring given that risk assessment and risk–benefit assessment are being widely promoted as managerial tools.

It is particularly important that these limitations are recognized when risk assessment is applied to the formulation of regulations and obliquely to the implementation and enforcement of such controls. Risk assessments should be viewed with deep scepticism unless all of the areas of uncertainty are explicitly

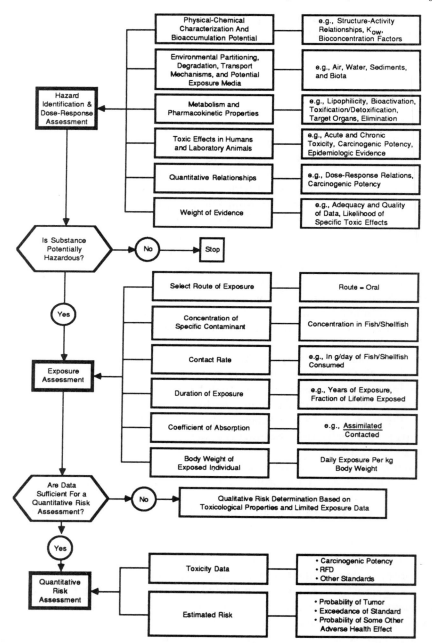

Figure 2 *A conceptual framework produced by the US EPA and reproduced from reference 9. This clearly shows the complex and intricate procedures necessary and the need for good quality analytical data in order to compute exposure factors. In addition, it must be noted that the framework is only capable of accommodating single chemicals for assessment purposes*

defined. In every case, risk assessment must be accompanied by penetrating reality checks at every stage of the process and particularly in the application of assessment results to the formulation of policy. Ultimately, it must be recognized, however,[48] that regulation is at best a compromise between what is achieved at the moment and what is desirable in the future. The best way to remove this element of compromise in the medium to long term is to accept that regulations merely represent resting places on the road to a goal of zero discharge of dangerous chemicals to the environment.

References

1 National Rivers Authority, 'Proposals for Statutory Water Quality Objectives', Report of the National Rivers Authority, Water Quality Series No. 5, NRA, Bristol, 1991, 100 pp.

2 P. A. Johnston, M. MacGarvin, and R. L. Stringer, 'Regulation of effluents and implications for environmental polity', *Water Sci. Technol.*, 1991, **24**(10), 19–27.

3 CEC, Commission of the European Communities, 'European Community Environment Legislation, Volume 7: Water', Brussels, 1992, 462 pp.

4 B. S. Shane, 'Introduction to Ecotoxicology', in 'Basic Environmental Toxicology', eds. L. G. Cockerham and B. S. Shane, CRC Press, Boca Raton, FL, 1994, ch. 1, pp. 3–10.

5 P. Edwards, 'Dangerous Substances in Water: A Practical Guide', Environmental Data Services Ltd, London, 1992, 64 pp.

6 Department of the Environment, 'Inputs of Dangerous Substances to Water: Proposals for a Unified System of Control (The Red List)', UK DoE, July 1988, 25 pp and Annexes.

7 Ed. G. W. Suter, 'Ecological Risk Assessment', Lewis Publishers, Boca Raton, FL, 1993.

8 J. Bol, H. J. M. Verhaar, C. J. Van Leeuwen, and J. L. M. Hermens, 'Predictions of the Aquatic Toxicity of High Production Volume Chemicals. Part A: Introduction and Methodology', Report No. 1993/9A, Ministerie van VROM, Den Haag, Netherlands, 1993.

9 Environmental Protection Agency, 'Assessing Human Health Risks from Chemically Contaminated Fish and Shellfish: A Guidance Manual', Report No. EPA-503/8/89-002, US Office of Marine and Estuarine Protection, Washington, DC, 1989.

10 International Commission on Radiological Protection, Recommendation of the ICRP Publication 26, Pergamon Press, New York, 1977.

11 Department of the Environment, 'Risk-benefit Analysis of Existing Substances', UK, February 1995, 69 pp.

12 Department of the Environment, 'This Common Inheritance: Britain's Environmental Strategy', HMSO, London, 1990.

13 Department of the Environment, 'Quality Status of the North Sea: A Report By the Scientific and Technical Working Group', UK, London, 1987, 88 pp.

14 North Sea Task Force, 'North Sea Quality Status Report 1993', Oslo and Paris Commissions, London, 1993, 132 pp.

15 J. C. Ellis, 'Determination of Pollutants in Effluents: Part B: Alternative Forms of Effluent Consents: Some Statistical Considerations', WRc Environment Report No: TR 235. WRc Environment, Medmenham, UK, 1986, 21 pp.

16 P. A. Johnston and R. L. Stringer, 'Analytical difficulties in the full characterisation of effluents', 1991, *Anal. Proc.* **28**, 249–250.

17 Royal Commission on Environmental Pollution, 16th Report: 'Freshwater Quality',

HMSO, London, 1992.

18 A. R. Agg and T. F. Zabel, 'Red-List substances: selection and monitoring', *J. Inst. Water Environ. Manage.*, 1990, **4**, 44–50.

19 A. J. Bale, 'Sampling considerations in the North Sea', *Annal. Proc.*, 1991, **28**, 264–265.

20 R. Hupkes, 'Pollution of the North Sea by Western European rivers', *Water Sci. Technol.*, 1991, **24**(10), 69–75.

21 W. C. Grogan, 'Input of Contaminants to the North Sea from the United Kingdom', Report to the UK Department of the Environment By the Institute of Offshore Engineering, Heriot-Watt University, Edinburgh, 1984.

22 W. Kuhn, G. Radach, and M. Kersten, 'Cadmium in the North Sea – a mass balance', *J. Mar. Syst.*, 1992, **3**, 209–224.

23 R. Chester, G. F. Bradshaw, C. J. Ottley, R. M. Harrison, J. L. Merrett, M. R. Preston, A. R. Rendell, M. M. Kane, and T. D. Jickells, 'The atmospheric distributions of trace metals, trace organics and nitrogen species over the North Sea', *Philos. Trans. R. Soc. London, Ser. A*, 1993, **343**, 543–556.

24 D. E. Wells, 'QUASIMEME – An Introduction', Quasimeme Bulletin No. 1, Scottish Office Agriculture and Fisheries Department, Aberdeen, March 1993.

25 D. E. Wells, W. P. Cofino, P. Quevauviller, and B. Griepink, 'Quality assurance of information in marine monitoring: A holistic approach', *Mar. Pollut. Bull.*, 1993, **26**(7), 368–375.

26 G. Rimkus, L. Rexilius, G. Heidemann, A. Vagts, and J. Hedderich, 'Results of an interlaboratory study on organochlorine compounds (PCB, DDT, DDE) in seal blubber (*Phoca vitulina*)', *Chemosphere*, 1993, **26**(6), 1099–1108.

27 Rijksinstitut voor Visserijonderzoek, 'Annual Report for 1992', RIVO, IJmuiden, Netherlands, 1993.

28 R. J. Law and J. E. Thain, 'Environmental monitoring and analysis in the North Sea under the North Sea Task Force', *Anal. Proc.*, 1991, **28**, 250–252.

29 R. J. Fryer and M. D. Nicholson, 'The power of a contaminant monitoring programme to detect linear trends and incidents', *ICES J. Mar. Sci.*, 1993, **50**, 161–168.

30 D. W. Evans, D. K. Dodoo, and P. J. Hanson, 'Trace element concentrations in fish livers: Implications of variations in fish size in pollution monitoring', *Mar. Pollut. Bull.*, 1983, **26**(6), 329–334.

31 P. A. Johnston, R. L. Stringer, R. Clayton, and R. J. Swindlehurst, 'Regulation of toxic chemicals in the North Sea: the need for an adequate control strategy', *North Sea Monitor*, 1994, **12**(2), 9–16.

32 L. A. Wallace, 'Human exposure to environmental pollutants: A decade of experience', *Clin. Exp. Allergy*, 1995, **25**, 4–9.

33 J. Cairns, 'The genesis of ecotoxicology', in 'Ecological Toxicity Testing: Scale Complexity and Relevance', eds. J. Cairns and B. R. Niederlehner, Lewis Publishers, Boca Raton, FL, 1995, ch. 1, pp. 1–10.

34 G. W. Suter, 'Endpoints of interest at different levels of biological organisation', in 'Ecological Toxicity Testing: Scale Complexity and Relevance', eds. J. Cairns and B. R. Niederlehner, Lewis Publishers, Boca Raton, FL, 1995, ch. 3, pp. 35–48.

35 G. Taubes, 'Epidemiology faces its limits: Special news report', *Science*, 1995, **269**, 164–169.

36 J. S. Gray, K. R. Clarke, R. M. Warwick, and G. Hobbs, 'Detection of initial effects of pollution on marine benthos: an example from the Ekofisk and Eldfisk oilfields, North Sea', *Mar. Col. Prog. Ser.*, 1990, **66**, 285–299.

37 F. Olsgard and J. S. Gray, 'A comprehensive analysis of the effects of offshore oil and gas exploration and production on the benthic communities of the Norwegian continental shelf', *Mar. Ecol. Prog. Ser.*, 1995, **122**(1–3), 277–306.

38 T. Colborn, 'The wildfire/human connection: modernizing risk decisions', *Environ. Health Perspect.*, 1994, **102**, (Suppl. 12), 55–59.
39 Danish Environmental Protection Agency, 'Male Reproductive Health and Environmental Chemicals with Estrogenic Effects', Danish Environmental Protection Agency, Copenhagen, 1995, 166 pp.
40 MRC, 'IEH Assessment on Environmental Oestrogens: Consequences to Human Health and Wildlife', UK Medical Research Council Institute for Environment and Health, University of Leicester, 1995, Assessment A1, 107 pp.
41 D. E. Wells, W. P. Cofino, P. Quevauviller, and B. Griepink, 'Quality Assurance of information in marine monitoring: A holistic approach', *Mar. Pollut. Bull.*, 1993, **26**(7), 368–375.
42 R. J. Law, T. W. Fileman, and P. Mattiessen, 'Phthalate esters and other industrial organic chemicals in the North and Irish Seas', *Water Sci. Technol.*, 1991, **24**(10), 127–134.
43 R. van Zoest and G. T. M. van Eck, 'Occurrence and behaviour of several groups of organic micropollutants in the Scheldt estuary', *Sci. Total Environ.*, 1991, **103**, 57–71.
44 L. S. McCarty and D. Mackay, 'Enhancing ecotoxicological modelling and assessment', *Environ. Sci. Technol.*, 1993, **27**(9), 1719–1728.
45 P. Matthiessen, J. E. Thain, R. J. Law, and T. W. Fileman, 'Attempts to assess the environmental hazard posed by the complex mixtures of organic chemicals in UK estuaries', *Mar. Pollut. Bull.*, 1993, **26**(2), 90–95.
46 P. A. Johnston and M. MacGarvin, '0–2000: Assimilating lessons from the past', Greenpeace 3rd North Sea Conference Report No: 28, Greenpeace International, Amsterdam, 1989, 32 pp.
47 V. J. Smith, R. J. Swindlehurst, P. A. Johnston, and A. D. Vethaak, 'Disturbance of host defence capability in the common shrimp *Crangon crangon* by exposure to harbour dredge spoils', *Aquat. Toxicol*, 1995, **32**, 45–58.
48 J. B. Sprague, 'Environmentally desirable approaches for regulating effluents from pulp mills', *Water Sci. Technol.*, 1991, **24**(3/4), 361–371.

Subject Index

Acceptable Daily Intake (ADI), 17, 23, 97
Air pollution
  area sources, 134
  benzene, 48ff, 100
  dust, 133
  Expert Panel on Air Quality Standards
    (EPAQS) (UK), 48, 50
  point sources, 134
Analytical data
  Quality Control and Quality Assurance
    Project (QUASIMEME), 227
As Low as Reasonably Achievable
    (ALARA) principle, 211
As Low as Reasonably Practicable
    (ALARP) principle, 103, 211

Base Set tests, 189
Benzene, 23, 27, 48, 100, 126
Benzo[a]pyrene, 47, 57, 60
Best Available Technique Not Entailing
    Excessive Cost (BATNEEC), 14, 83,
    86, 90, 103, 108
Best Environmental Option (BEO), 88
Best Practical Environmental Option
    (BPEO), 8, 85, 87, 90, 108
Bioconcentration factor (BCF), 122
Biotransfer factor (BTF), 122
British Library, 28, 151
  Environmental Information Service
    (EIS), 147, 151ff
Butadiene, 50

Carcinogenesis, 54
  glossary of terms, 64
Carcinogenic potency, 61
Carcinogens
  benzene, 23, 27, 48, 49
  benzo[a]pyrene, 57
  butadiene, 50
  environmental, 47, 54

formaldehyde, 55
genotoxic, 21, 44, 54, 59, 97
non-genotoxic, 23, 59
propylene oxide, 56
risk assessment, 22, 45
synergism, 63
Centre for Integrated Environmental Risk
    Assessment (CIERA) (UK), 8, 14
Chemicals of Potential Concern (COPC),
    120
Chronic Daily Intake (CDI), 26, 27, 138
Closed landfill sites, 13
Combustion emissions, 6, 134
Committee on Carcinogenicity of
    Chemicals in Food, Consumer
    Products, and the Environment (UK),
    46, 100
Contaminated land, 7, 93, 106, 128
  ALARP principle, 103, 211
  assessment, 100
  dioxins, 10
  dust emissions, 133
  Dutch ABC list, 107
  guidelines, 101, 106
  human health risks, 94, 120
  redevelopment, 103
  vapour emissions, 129, 130, 132
Contaminated Land Exposure Assessment
    (CLEA), 8, 100

Department of the Environment (DoE)
    (UK), 6, 7, 9, 11, 48, 95, 106, 186, 237
Dichloromethane, see Methylene chloride
Dictionary of Substances and their Effects
    (DOSE), 156, 181
Dioxins, 10, 22

Environmental assessment,
  European Union, 192, 203
  UK, 86